THE

GSH
PHENOMENON

T H E

GSH
PHENOMENON

NATURE'S MOST POWERFUL ANTIOXIDANT AND HEALING AGENT

Alan H. Pressman, D.C., Ph.D., C.C.N.

with Sheila Buff

Foreword by Richard A. Passwater, Ph.D.

ST. MARTIN'S PRESS ≈ NEW YORK

PRODUCED BY THE PHILIP LIEF GROUP, INC.

This book is meant to educate and should not be used as an alternative to proper medical care. No supplements or vitamins mentioned herein should be taken without qualified medical consultation and approval. The authors have exerted every effort to ensure that the information presented is accurate up to the time of publication. However, in light of ongoing research and the constant flow of information, it is possible that new findings may invalidate some of the data presented here.

Design by Helene Wald Berinsky

Library of Congress Cataloging-in-Publication Data

Pressman, Alan H.
    The GSH phenomenon : nature's most powerful antioxidant and healing agent / Alan H. Pressman, with Sheila Buff; foreword by Richard A. Passwater, Ph.D.—1st ed.
        p. cm.
    "Produced by the Philip Lief Group"
    Includes bibliographical references and index.
    ISBN 0-312-15135-7
    1. Glutathione—Popular works. I. Buff, Sheila. II. Title.
QP552.G58P74   1997
612'.015756—dc20   96-17980
                                                                                    CIP

First Edition: February 1997

10   9   8   7   6   5   4   3   2   1

To my son, Corey,
and my daughter, Meghan,
because they are my life

C O N T E N T S

Foreword by Richard A. Passwater   ix

1. Amino Acids, Glutathione, and Optimal Health   1

2. Glutathione and Liver Detoxification   21

3. Glutathione and Environmental Illness   55

4. Glutathione and Cancer Prevention   79

5. Glutathione and Aging   93

6. Glutathione and Your Immune System   117

7. Glutathione and Your Eyes   139

8. Glutathione and Other Ailments   157

9. Eating Your Glutathione   175

Appendix A: Testing Laboratories and
      Holistic Health Organizations   205

Appendix B: Sources of Supplements   209

Glossary   213

Further Reading   215

Index   217

Your body is under constant attack, and your health is at stake! The enemy? Toxins: environmental agents in our food, our water, our air—and our bodies. The foot soldiers in the battle against toxins are antioxidant molecules in your cells called glutathione. This all-natural intracellular chemical found in the bloodstream is known as a tripeptide because it is composed of three essential amino acids: cysteine, glutamic acid, and glycine.

Reports of how antioxidants protect us against cancer, heart disease, and nearly eighty other diseases appear regularly in the media. Vitamin C is the major antioxidant in the blood; vitamin E is the vital antioxidant in cell membranes and lipoproteins. However, in the cell interior—where the real toxin war is being waged—the most important antioxidant is glutathione. But rarely is glutathione mentioned, in spite of its central role.

I have researched glutathione and other sulfur-containing compounds since 1959 and have incorporated them into age-retarding and cancer-prevention formulas since 1970. Today, there is no question about the connection between glutathione and good health. One example is found in the treatment of AIDS: When a patient with AIDS has a lowered glutathione level, the immune system falters.

Fortunately, your cellular glutathione level can be improved by diets rich in glutathione and its elements, as well as by food supplements that boost your glutathione level. In this book, Dr. Alan

Pressman explains how to use glutathione and its partners to protect and improve your health. His clinical experience is invaluable
in translating the scientific knowledge into practical advice. If you
follow his advice on incorporating glutathione into your lifestyle,
you will be amazed at the improvement in your health and energy.
It may even save your life. Don't wait until the toxins claim terri-
tory in your body and cause disease. Get a head start today battling
them, thanks to Dr. Pressman.

RICHARD A. PASSWATER, PH.D., author of
Cancer Prevention and Its Nutritional Therapies

THE

GSH
PHENOMENON

CHAPTER 1

AMINO ACIDS, GLUTATHIONE, AND OPTIMAL HEALTH

Daniel was getting progressively weaker. His joints and muscles ached, he felt mentally confused, and he was always tired. His work as an attorney was suffering badly because he just couldn't concentrate. His doctor listened sympathetically, but examinations and lab tests found nothing wrong. Based on his medical results, Daniel should have been the picture of health. His doctor thought Daniel was suffering from depression and prescribed antidepressant medication. Daniel took the pills, but they only made him feel worse and more exhausted than ever.

In desperation, Daniel decided to consult a nutritionist—me. A complete nutritional evaluation revealed that Daniel's liver was functioning at only half its normal capacity. Daniel's inefficient liver was letting wastes build up in his body, like a toxic storage dump. No wonder he was tired all the time.

I worked with Daniel to improve his liver function through nutrition. The mainstay of his program was a natural substance called glutathione (gloo-ta-*thigh*-on), along with some other nutrients that improve the efficacy of glutathione. His improvement was dramatic. Within ninety days, Daniel was a new man, restored to his previous energy and mental ability.

To understand why adding a few capsules containing a simple brown powder to his daily diet helped Daniel so much—and how it can help you too—you need to understand a little bit more about free radicals, antioxidants, and glutathione. Once you understand them, you'll recognize that glutathione is *your* best line

1

of natural defense against environmental toxins, cancer, heart disease, cataracts, and many other health problems.

Your body naturally produces glutathione in abundance—it is essential to life itself. Glutathione circulates constantly throughout your body, neutralizing free radicals and removing dangerous waste products and toxins from your system. When your glutathione level is high, your overall health is high. You feel good and you look good. You fight off minor illnesses quickly, you have plenty of energy, and you feel mentally and physically alert.

If your glutathione level is high, you're at an optimal level of good health, for now and for the future. Glutathione naturally protects you from the long-term dangers of free radicals, metabolic wastes, and environmental poisons. Since those dangers include cancer, heart disease, premature aging, autoimmune diseases, and chronic illnesses such as asthma, keeping your glutathione level high helps you stay healthy and active.

Daniel is just one of the many patients whom I have helped using glutathione. They have suffered from a wide range of serious ailments, and they've all improved rapidly after adding glutathione to their diets.

You too can benefit from glutathione, quickly and easily. There's nothing mysterious about it. Glutathione is a natural protein found in fresh fruits and vegetables. To achieve better health with glutathione, you don't have to go on a weird diet, give up your favorite foods, or take powerful drugs. All you need to do is make sure your daily diet contains enough nutrients to make glutathione; you also need to have some important vitamin and mineral cofactors. And you can be sure you're getting enough of these essential nutrients simply by taking inexpensive supplements found in any drugstore or health food store. It's safe, it's simple, and it works.

The chapters that follow will explore the many ways glutathione functions in your body to help prevent and heal illness. This chapter explains free radicals, antioxidants, and the role of glutathione in the body. Chapter 2 discusses the essential role of glutathione in good liver function—and how good liver function is central to your overall health. Chapter 3 explores how glutathione can help prevent and treat environmental illness resulting from toxic exposure. Chapter 4 explains how glutathione can help prevent cancer. Chapter 5 tells how glutathione can promote longevity and protect against heart disease, arthritis, Alzheimer's disease, and other problems of older adults, and chapter 6 explains the central role of glutathione in boosting immune system strength and fighting autoimmune diseases, such as lupus. Chapter 7 focuses on how glutathione can help keep your eyes healthy, and chapter 8 explains how glutathione can help prevent or treat a number of other common diseases, including asthma, bronchitis, psoriasis, and chronic fatigue syndrome. And chapter 9, on diet and glutathione, wraps up the book and shows you how to keep your glutathione level high through the foods you eat. This chapter includes some of my favorite recipes.

Glutathione is central to continuing good health. The underlying concepts are easy to grasp, and the potential benefits are huge.

THE ANTIOXIDANT AMINO ACIDS

You've probably heard the words *free radical* and *antioxidant* a lot lately. Researchers have been discovering that free radicals in your body are the underlying cause of a lot of serious illness, including heart disease, cancer, cataracts, and a host of other problems. Fortunately, they've also discovered that antioxidants can counteract free radicals and actually help prevent disease. But what exactly is a free radical? And why do you need antioxidants so much?

As your body goes through its normal metabolic process of converting food to energy to fuel your cells, it creates free radicals as a byproduct. A free radical is an unstable oxygen-containing molecule with an unpaired electron in its outer orbit. That single electron desperately wants to attract another electron to it to make a stable pair. In order to do that, the molecule crashes around within your cells until it can grab onto another electron. If it seizes an electron from a cell's internal structure, its delicate membrane, or its nucleus, the cell is damaged. And if the cell is damaged severely enough or often enough, it stops working efficiently, mutates, or even dies. In addition, the molecule that has been robbed of its electron itself becomes a free radical, starting a dangerous cascade that can be stopped only if your body can rapidly find a way to quench the free radicals and break the cycle of cell destruction.

So far, scientists have identified two kinds of destructive free radicals: the hydroxyl radical and the superoxide radical. Two other kinds of oxygen-containing molecules, hydrogen peroxide and singlet oxygen, are just as dangerous.

Fortunately, your body has its own natural way to deal with all four of these renegade molecules.

The same energy-producing process that creates free radicals also creates their antidote—the antioxidant enzymes glutathione peroxidase, catalase, and superoxide dismutase (SOD). These enzymes tackle the free radicals in your cells, neutralizing the destructive free electrons in the oxygen molecules almost as soon as they are formed. The neutralized free radicals can then be eliminated easily from the body. Collectively, antioxidant enzymes help maintain a healthy balance between reduction (gaining an electron) and oxidation (giving up an electron). Scientists refer to this ongoing metabolic process as redox (reduction–oxidation) operations.

Of all the antioxidant substances your body makes, the most important and abundant is glutathione. This incredibly powerful natural antioxidant is your built-in defense against the harmful effects of free radicals. Glutathione also plays a crucial role in removing metabolic wastes and in finding and eliminating toxic substances, such as heavy metals and other environmental poisons, from your body. Glutathione helps keep your cell membranes strong and helps transport vital amino acids into your cells.

Glutathione was discovered in 1888, but it wasn't until the 1920s and 1930s that researchers began to unfold its mysteries. Initially, most of the research concentrated on the eye, especially on the lens. The research was very valuable—glutathione deficiency is directly related to eye problems such as cataracts and macular degeneration (see chapter 7). By the mid-1980s, however, researchers had learned that glutathione's impact on the body goes far beyond the eyes. In fact, it is vital for keeping your body's systems in good working order. A lack of glutathione can lead to disruption of your body's ability to detoxify itself, which in turn can lead to cancer, heart disease, joint problems, and serious imbalances in your immune, endocrine, and nervous systems.

If you're basically healthy, your body's metabolic processes maintain a natural redox balance, rapidly quenching unavoidable free radicals with antioxidants. No matter how healthy you are, however, the redox balance can easily get disrupted. Whenever you're under stress, sick, fighting an infection or inflammation, or healing from injury, you create more free radicals. Free radicals are also created when your body is bombarded by the almost unavoidable contaminants that surround us: cigarette smoke, alcohol, ultraviolet light, heavy metals such as mercury, air pollution, pesticides, food additives, and other environmental poisons. And although glutathione circulates throughout your system, constantly on the lookout for free radicals and toxic wastes, the balance can easily tip in favor of the renegades.

FIGHTING BACK WITH GLUTATHIONE

Your body needs superoxide dismutase to recycle the oxygen in the superoxide radical. It also needs catalase to recycle the oxygen in hydrogen peroxide. But most of all, your body needs glutathione, because it corrals all kinds of free radicals before they can attack the delicate and critically important cell membrane. Known as lipid peroxidation, this free-radical destruction of the crucial fatty acids that form the cell membrane quickly leads to destruction of the cell.

Glutathione in your body exists in two forms: the reduced form (GSH) and the oxidized form (GSSG). Several additional enzymes are necessary for glutathione to function effectively. The most important is glutathione peroxidase, which works with glutathione as a catalyst (an agent that speeds up a reaction) to capture and neutralize free radicals, especially toxic hydrogen peroxide. Glutathione reductase is a support enzyme that recycles glutathione and other antioxidants after they have neutralized free radicals. This recycling helps remove the captured free radicals from the body via the usual excretory pathways. It also frees up the glutathione to go out again and capture even more free radicals.

Glutathione is a tripeptide, which means it is made from the linked molecules of three nonessential amino acids: cysteine, glutamic acid, and glycine. Your body gets some glutathione from food, but it manufactures most of the glutathione it needs from these three building blocks. In turn, the three building blocks are made from different combinations of essential amino acids.

Of course, glutathione metabolism is actually a lot more complicated. In general, whenever glutathione is mentioned here, it refers to the reduced form GSH and not the various catalyst enzymes.

The abbreviation for glutathione is GSH. The *-SH* suffix indicates that reduced glutathione contains a compound called sulfhydryl. Your body gets the sulfhydryl it needs to make GSH from the precursor amino acids cysteine and methionine, which contain sulfur.

In order to maintain an optimal level of glutathione production, your body must maintain an optimal level of the amino acid raw materials. You also have to maintain an optimal level of the cofactors that help glutathione do its work: the minerals selenium and zinc, lipoic acid, and vitamin B_2 (riboflavin). Selenium is particularly important because it is a vital component of the enzyme glutathione peroxidase.

WHAT ARE AMINO ACIDS?

The twenty different amino acids are the building blocks of all life on earth. They are found in all living plants and animals, from the simplest one-celled organisms to the magnificently complex human body. Each amino acid is made from atoms of just four or five very common elements—oxygen, hydrogen, nitrogen, carbon, and sometimes sulfur—in different combinations.

When different amino acids are joined, they form chains of molecules called proteins. Each protein is created as it is needed for a specific function as dictated by the DNA in your genes. Among other things, your body uses proteins for growth; to maintain and repair cells; to defend your body against infection; as chemical messengers, such as enzymes, neurotransmitters, and hormones; and as antioxidants. So far, researchers have identified

Figure 1.1

THE AMINO ACIDS

Essential	Nonessential
Histidine	Alanine
Isoleucine	Arginine
Leucine	Asparagine
Lysine	Aspartic acid
Methionine	Carnitine
	Cysteine
Phenylalanine	Glutamic acid
Taurine	Glutamine
Threonine	Glycine
Tryptophan	Proline
Valine	Serine
	Tyrosine

more than fifty thousand distinct kinds of proteins in the human body—all made by your cells just from different combinations of the twenty basic amino acids.

Scientists divide the amino acids into two groups: essential and nonessential. Essential amino acids are the ten aminos you can get *only* from the foods you eat. (Taurine is not considered essential beyond infancy by some researchers, although more and more evidence suggests that it is important to adults.) Nonessential amino acids are those that your body can manufacture by recombining elements from the essential amino acids. Nonessential here doesn't mean unnecessary, however—you need *all* the amino acids to live. (See Figure 1.1 for a list of the amino acids.)

Whenever you eat protein, whether plant or animal, you're actually eating long chains of amino acids. Your digestive system breaks those long chains down into their components and absorbs them into your bloodstream. From there, they are distributed to the cells of your body, to be reassembled into whatever protein your body happens to need from moment to moment. After the protein has been made and has performed its function, it is broken down again by enzymes; most of the amino acids are then available to be reutilized in another protein. The byproducts of this constant recycling are mostly nitrogen-based wastes, such as uric acid, that are excreted from your body through your urine, bile, sweat, and other pathways. So, even though your body reuses its amino acids very efficiently, some are broken down and excreted as byproducts, so you still need a continuous new supply. And because your body can't store amino acids well (excess amounts are mostly just excreted), you have to maintain a good balance every day to sustain your body's good health.

GETTING YOUR AMINOS

At one time, I thought that good nutrition with the recommended levels of plant or animal protein was all you needed to give your body the right amounts of amino acids to stay well. After years of following glutathione research and years of seeing how supplemental GSH and amino acids have helped my patients, I've changed my mind. Good nutrition is only the starting point. Supplementing your diet with glutathione and its building blocks, along with supplements of the needed cofactors, will take you beyond low-level wellness into the realm of optimal health and long life.

Here's why. First, we can't always eat the right foods. Often we're just too rushed or tired to select and prepare a healthful meal. Second, whenever you're stressed, exercising heavily, sick, or injured, you need additional antioxidants to combat the extra free radicals your body produces. Under adverse conditions, your normal diet may not provide enough building blocks and cofactors to produce enough glutathione to meet the free radical challenge. Even worse, if you're sick or stressed out, you're probably not eating right, which only makes matters worse by giving those free radicals the upper hand.

Most important, all of us, no matter how healthy, fit, and nutrition-conscious, are constantly being bombarded with environmental contaminants that produce free radicals. We need to boost our level of antioxidant GSH just for simple self-protection from the world around us. And to do that, we need to take in more GSH building blocks and cofactors than we can reasonably consume through food alone.

Figure 1.2

SOME AMINO ACID FUNCTIONS

Amino Acid	Function
Arginine	Stimulates immune system; vital for healthy connective tissue and skin; helps wound healing and muscle growth
Cysteine	Removes heavy metals and free radicals; contains sulfur
Glutamine	Nourishes small intestine; helps wound healing; helps remove waste products; strengthens immune system
Glutamic acid	Needed to synthesize glutamine in the brain; needed to synthesize neurotransmitters; can raise blood sugar levels
Glycine	Necessary to build collagen and connective tissues; helps remove waste products
Histidine	Needed to synthesize neurotransmitters; aids immune system; regulates allergic responses; removes heavy metals; aids in production of stomach acid
Isoleucine	Aids wound healing and muscle growth
Leucine	Aids wound healing and muscle growth
Lysine	Needed to build connective tissues, collagen, and bone; assists transport of calcium throughout the body
Methionine	Provides sulfur; removes heavy metals and quenches free radicals
Phenylalanine	Needed to synthesize insulin, adrenaline, neurotransmitters, other enzymes; helps build collagen and other connective tissues
Proline	Necessary to build collagen and connective tissues
Taurine	Aids in regulation of central nervous system; essential for babies
Tryptophan	Needed to utilize B vitamins; needed to synthesize the essential neurotransmitter serotonin
Valine	Aids wound healing and muscle growth

BOOSTING YOUR GSH LEVEL

You can boost your GSH level in different ways: taking supplements of GSH itself, or taking supplements of GSH amino acid building blocks and cofactors, or both. Which route you choose will depend to a degree on your overall health and your personal preferences.

Taking GSH supplements is a little controversial. The argument against supplements is that even though GSH is a small tripeptide, its molecules are too large to be absorbed into the body through the walls of your small intestine. Instead, the molecules are broken down in the gut into their component amino acids, which are then absorbed. However, numerous studies suggest that the body can absorb tripeptides such as GSH through the gut. Other studies indicate that the plasma level of GSH rises temporarily in some subjects when supplements are taken, but only if the subjects had a low glutathione level to begin with. However, for subjects with a normal glutathione level, some of the GSH is broken down by the normal digestive processes and doesn't reach the bloodstream in tripeptide form. Studies aside, on a purely practical level I have repeatedly seen my patients improve once they started taking GSH supplements.

If you are in good health with no digestive problems or chronic illnesses, you can probably benefit from taking GSH supplements and GSH building blocks along with a nutritious diet high in GSH-containing fresh fruits and vegetables. Foods such as avocados, asparagus, grapefruit, potatoes, strawberries, and watermelon are naturally high in glutathione. Other foods, including parsley, spinach, and cruciferous vegetables such as broccoli, contain substances that naturally stimulate your body to produce more glutathione. In addition, you should take the supplemental cofactors lipoic acid, selenium, and riboflavin (vitamin B_2). Your body produces its own lipoic acid, but you can boost your level by taking supplements; brewer's yeast also contains lipoic acid. Selenium is found in whole grains, organ meats, lean meat and poultry (especially turkey), Brazil nuts, and seafood. Good dietary sources of riboflavin include milk, salmon, avocado, dark green leafy vegetables, asparagus, broccoli, and brussels sprouts.

Because not everyone seems to absorb GSH well, you might not feel any positive effects from taking supplements alone. In that case, I recommend taking supplements of the component amino acids and cofactors. If your overall health is impaired, if you have any sort of digestive difficulties, or if you have a chronic illness, you should definitely consider supplements of glutathione and all its cofactors. Specifically, I recommend taking supplemental cysteine in the form of N-acetyl-L-cysteine capsules. Called NAC for short, it is a natural derivative of the nonessential amino acid cysteine. For reasons that scientists are still researching, your body absorbs NAC better than unmodified cysteine. Of the three amino acids your body needs to manufacture glutathione—cysteine, glycine, and glutamic acid—cysteine is the most important, since only cysteine contains the crucial sulfur molecule. This means that the level of cysteine in your system is the limiting factor in how fast you can produce glutathione and how much of it you can make. By taking a NAC supplement, you guarantee that your body has enough cysteine to make the glutathione you need, when you need it. Cysteine is found naturally in many foods, particularly in whole grains, beans, eggs, and lean meats; high-quality, inexpensive NAC capsules are available at health food stores. According to researcher Carl Germano, RD, CNS, LDN, of the Solgar Company (a leading maker of all-natural supplements), NAC is best made by a fermentation process utilizing friendly bacteria.

In addition, I recommend taking the other two building blocks of glutathione in the form of supplemental glycine and glutamine. (Your body easily converts the glutamine to glutamic acid.) For glycine, I suggest 250 to 500 mg daily in the form of capsules (see Appendix A). For glutamine, I suggest the powder form: Stir one teaspoon (about 1,500 mg) into a glass of juice or water; take one to three teaspoons a day.

Compare the amino acid dosages I recommend in Figure 1.4 to the recommended dietary allowances chart in Figure 1.3; my dosages are considerably higher. This is because I, along with many other researchers, nutritionists, and holistic health practitioners, believe that the RDA charts present only the bare minimum amounts needed for sustaining life, rather than the amounts needed for optimal good health.

Figure 1.3

RECOMMENDED DIETARY ALLOWANCES FOR ESSENTIAL AMINO ACIDS

Amino Acid	RDA in milligrams per kilogram (mg/kg)
Histidine	8–12
Isoleucine	10
Leucine	14
Lysine	12
Methionine plus cystine	13
Phenylalanine plus tyrosine	14
Threonine	7
Tryptophan	3.5
Valine	10

The recommended dietary allowances given here are for adults. To convert pounds to kilograms, multiply your weight in pounds by 2.2.

Source: National Research Council, "Protein and Amino Acids," in *Recommended Dietary Allowances*, 10th edition (1989).

COFACTORS FOR GLUTATHIONE

Lipoic acid is another important cofactor for glutathione synthesis in the body. An antioxidant by itself, lipoic acid is also an important coenzyme that helps recycle the other antioxidants in your body. It is vital for converting glutathione back and forth from its oxidized to its reduced form. Lipoic acid also aids in recycling the antioxidant vitamins C and E and helps preserve nerve cell growth and liver cell function. If you take amino acid supplements without taking lipoic acid, you're not providing everything your body needs to synthesize GSH and utilize its antioxidants effectively. According to biochemist Richard A. Passwater, a leading antioxidant researcher, lipoic acid is a coenzyme necessary for producing energy in your cells, a powerful antioxidant, and is also a chelator (a substance that binds to metal) that helps remove toxic metals such as lead and cadmium from your body.

I generally recommend capsules of alpha-lipoic acid, since

this is the most active form of lipoic acid available. The best source of single alpha-lipoic acid is the Solgar Company. I generally recommend taking about 200 mg daily to support GSH synthesis.

Two other cofactors are necessary for complete GSH supplementation: the mineral selenium and vitamin B_2 (riboflavin). Selenium is a necessary component of glutathione peroxidase, an essential catalyst enzyme for glutathione. Selenium also works synergistically with vitamin E—each improves the effectiveness of the other, and both improve the effectiveness of glutathione. Riboflavin is needed to help your body combine amino acids, including glutathione, into proteins.

In addition to taking your aminos and cofactors, I also recommend supplements of vitamins C and E, mixed carotenes, and zinc, necessary and valuable sources of additional protection against free radicals. It's particularly important to get enough vitamin E, since it is also a needed cofactor for glutathione. In addition, vitamin E acts synergistically with selenium.

As you may have noticed, vitamin A is not on the list of suggested supplements. Your small intestine and liver manufacture all the vitamin A your body needs from the carotenes you ingest. Carotenes are fat-soluble, natural hydrocarbons found in many vegetables and fruits, especially those that are orange, yellow, or red in color, such as carrots and sweet potatoes (although carotenes are also abundant in dark green leafy vegetables, such as kale). Of the fifty to sixty different carotenes in foods, only fourteen are actually absorbed into your system. Four of these, including beta carotene, are used to make vitamin A. The others, which include lycopene (abundant in tomatoes and watermelons), lutein (found in egg yolks), and zeaxanthin (found in egg yolks and corn), are valuable antioxidants. Adequate levels of zinc are needed for your body to manufacture vitamin A from carotenes.

Many of my patients prefer to take a single capsule rather than a confusing number of different supplements. I recommend a formula called GSH 250 Master Glutathione Formula, made by Douglas Labs (see Appendix B). The ingredients in each capsule are listed on the next page.

NAC	250 mg
Lipoic acid	200 mg
GSH	250 mg
Selenium	25 mcg
Riboflavin (vitamin B$_2$)	25 mg
Cysteine	5 mg
Glutamine	5 mg
Glycine	5 mg

The ingredients are in a base of spinach and parsley—natural food sources that stimulate your body to manufacture glutathione. If you feel you don't need such a high level of amino acid supplementation, I recommend Glutaplex, also from Douglas Labs. This is essentially the same formula as GSH 250 Master Glutathione Formula, but contains only 50 mg each of NAC, lipoic acid, and GSH. If you routinely eat foods high in glutathione or glutathione cofactors, Glutaplex may be all you need.

Whichever brand you choose, I generally suggest taking three capsules daily for maintaining basic good health. Since you can absorb the ingredients best on an empty stomach, take one tablet about an hour before eating. By spreading your dosage out over the day, you ensure that your body always has enough of the cofactors it needs to manufacture and utilize glutathione.

Figure 1.4 lists the total daily minimum amino acids, cofactors, vitamins, and minerals I recommend for maintaining good health. You don't need to buy separate containers of every single supplement on the list. It's much simpler and easier to purchase combined formulas. In addition to your glutathione formula, I suggest taking two capsules daily (one in the morning and one at night) of a high-quality multi supplement such as those made by Solgar, Douglas Labs, or Twin Labs.

Figure 1.4

DAILY SUPPLEMENTS FOR MAINTAINING BASIC HEALTH

This recommendation is for healthy adults who have no chronic conditions, acute illnesses, or significant allergies.

Amino Acids and Cofactors:

NAC	250 mg
Lipoic acid	200 mg
GSH	250 mg
Glutamine	1,500 to 5,000 mg
Glycine	250 to 500 mg

Vitamins:

Mixed carotenes	up to 15,000 IU
Vitamin C	up to 1,000 mg
Vitamin D	50 to 100 IU
Vitamin E	up to 400 IU
B vitamins complex:	

thiamin (B_1)	up to 100 mg
riboflavin (B_2)	up to 50 mg
niacin	up to 150 mg
pantothenic acid	up to 500 mg
pyridoxine	up to 50 mg
B_{12}	up to 500 mcg
folic acid	up to 800 mcg
biotin	up to 400 mcg

Minerals:

Calcium	1,000 to 1,500 mg
Magnesium	up to 500 mg
Potassium	up to 100 mg
Selenium	up to 250 mcg
Zinc	up to 30 mg

FORMS OF AMINO ACIDS

With amino acid supplements, always read the label carefully. Be sure to purchase only free-form aminos.* This means that the formulation contains just those particular aminos in their pure form. Because free-form aminos are already in their simplest form, they are absorbed into your body immediately. If the label doesn't say free-form, the aminos you want are probably present only as part of a protein chain made up of a combination of amino acids. Your digestive system will have to break down the chain in order to release the amino acids.

Most health food stores carry amino acid formulas that contain all the essential and nonessential amino acids. Read the labels carefully. Some of these formulas are nothing but protein powders with some added free-form amino acids. They're high in calories because they're designed for weight gain and building body mass, not for antioxidant protection. Even the complete formulations that contain only free-form amino acids usually don't contain enough cysteine (NAC) relative to the other amino acids, and they don't contain lipoic acid or the other needed cofactors.

If you eat a good, nutritious diet most of the time and take the GSH-boosting supplements discussed above (especially GSH Complex), your diet will offer an ample supply of the other amino acids you need. And if you have certain rather rare health conditions, taking supplements of every amino acid might not be a good idea (see the cautions below). On the other hand, taking a complete amino acid supplement in addition to supplemental GSH or NAC, along with lipoic acid and the other cofactors, will almost certainly be helpful.

Most people take their amino supplements in convenient cap-

---

*Free-form amino acids are available in two chemically distinct forms: the L form and the D form. The letters refer to the chemical structure of the amino acid. Basically, the two forms are mirror images of each other. L form aminos are "left-handed" and are all derived from protein sources, while D form aminos are "right-handed" short peptides produced by some bacteria. Humans can't metabolize the D forms; some D form amino acids have also been shown to depress the immune system and inhibit the effect of antibiotics. For these and other reasons, the federal Food and Drug Administration bans the sale of D form aminos. DL-phenylalanine and DL-methionine are unusual exceptions that contain both forms.

sule form. If you prefer to purchase powdered amino acids, simply place the powder on your tongue and wash it down with a few swallows of liquid. Aminos don't dissolve in water, so you'll have some difficulty stirring them into a beverage. Never add free-form amino acids to hot foods or use them in cooking—the heat alters their structure (denatures them) and makes them ineffective. This is especially true of glutamine.

Spread your amino acid and cofactor supplements out evenly across the day—this helps maximize your uptake and maintain constant levels in your system. Try to take your supplements about an hour before eating. Otherwise, the nutrients in the supplements will compete with the nutrients in your meal for absorption through the intestines. If you do take your supplements with a meal, go easy on proteins to prevent competition. Eat more carbohydrates instead.

CAUTIONS ABOUT
AMINO ACID SUPPLEMENTS

If you have kidney disease (except for occasional kidney stones), adding extra amino acids to your diet could cause problems. Be sure to discuss amino acid supplementation with your doctor or nutritionist.

Although glutathione can be extremely helpful in treating liver problems, it must be used very cautiously in very severe or end-stage cases. If you have severe liver disease, particularly from cirrhosis or Reye's syndrome, discuss amino acid supplementation with your doctor. For an extensive discussion of how glutathione can help your liver, see chapter 2.

People with seizure disorders such as epilepsy should avoid cysteine in large doses, although NAC in moderation is probably safe. If you have a seizure disorder, discuss amino acid and GSH supplements with your doctor.

In large doses (above 3,000 mg), lysine can interfere with insulin production, so don't take supplements of this amino acid if you have diabetes or blood sugar problems. On the other hand, lysine is often very helpful for treating and preventing outbreaks of herpes and cold sores.

Some rare genetic conditions are worsened by amino acids. People with phenylketonuria, an inherited inability to make the enzyme that converts the essential amino acid phenylalanine to the nonessential amino acid tyrosine, must avoid phenylalanine in all forms.

If you take a monoamine oxidase (MAO) inhibitor drug such as Nardil or Marplan (drugs often used to treat depression and anxiety) you should also avoid foods and amino acid supplements containing phenylalanine, tryptophan, and especially tyrosine. If you'd like to try supplements of the other amino acids, talk to your doctor first. And if you'd like to stop taking powerful drugs for depression, see chapter 3.

If you have a malabsorption problem, taking supplements of GSH or amino acids could be problematic. Before you start taking maintenance doses, read chapter 2 and go through the checklist in Figure 2.2. If you have more than two of the symptoms listed more often than three times a week, or if you have food allergies, you may have an imbalance in your intestinal flora that is causing malabsorption. Until the imbalance is corrected, your benefit from the supplements will be minimal. We'll discuss treating intestinal overgrowth and malabsorption in chapter 2.

As of now no reliable blood test for glutathione level is available. If you'd like to know more about your body's amino acid levels, talk to your physician or nutritionist about having a quantitative urinary amino acid screening. This test, currently available only through a few laboratories (see Appendix A), requires you to collect your complete urinary output over a twenty-four-hour period. Using a sophisticated liquid chromatograph, the laboratory analyzes the urine for traces of a number of different amino acids and compares your results to the norm. In the next chapter, we'll discuss some additional tests that can also give you a detailed picture of your glutathione level. I'd like to emphasize that while lab tests can provide you with valuable information about your metabolic pathways, they are *not* needed before starting a program of GSH or amino acid supplementation.

STICKING WITH IT

Although I have seen some near-miraculous improvements in patients who started taking supplemental GSH, NAC, and lipoic acid, changes don't happen overnight. Start your supplement program with the basic daily maintenance levels in Figure 1.4. By the end of three or four weeks, you will probably notice that some of those nagging minor aches and pains have vanished. You'll probably be feeling better overall, healthier and more energetic.

If you haven't noticed any real change for the better after a month, try increasing your daily maintenance level by one dose. If you still don't feel any improvement after another month, add another dose, up to a maximum of four doses a day, or 1,000 mg of GSH. You can't really overdose on amino acids or glutathione, but if you're still not experiencing any overall improvement by that point, it's time to visit your doctor or your nutritionist again to discuss your diet and check for absorption problems.

As you continue your maintenance program over the coming months, you'll probably also notice that you look and feel generally healthier. Your skin, hair, and nails will look better, you'll have fewer colds and minor illnesses, and minor injuries will heal faster. You'll have more energy and feel more positive. You'll bounce back faster from stress, illness, hard exercise, or a stretch of overwork.

But the most important benefits of glutathione could well be the things that *won't* happen to you. You're protecting yourself against neurotoxins and reducing your chances of cancer, heart disease, cataracts, premature aging, autoimmune illness, and other serious ailments. Stick with it—you'll be glad you did.

REFERENCES

Armstrong, D., R. S. Sohal, et al. *Free Radicals in Molecular Biology, Aging, and Disease* (New York: Raven Press, 1984).

Diplock, Anthony T. "Antioxidant Nutrients and Disease Prevention: An Overview," *American Journal of Clinical Nutrition*, vol. 53 (supplement), pp. 189S–93S, 1991.

Forman Henry Jay, Rui-Ming Liu, and Michael Ming Shi. "Glutathione Synthesis in Oxidative Stress," in *Biothiols in Health and Disease,* Lester Packer and Enrique Cadenas, eds. (New York: M. Dekker, 1995).

Lonsdale, Derrick. "Free Oxygen Radicals and Disease," *1986: A Year in Nutritional Medicine Monographs* (New Canaan, Conn.: Keats Publishing, 1986).

Meister, Alton. "Strategies for Increasing Cellular Glutathione," in *Biothiols in Health and Disease,* Lester Packer and Enrique Cadenas, eds. (New York: M. Dekker, 1995).

National Research Council. "Protein and Amino Acids," *Recommended Dietary Allowances,* 10th edition (Washington, D.C.: National Academy Press, 1989).

Null, Gary. "The Antioxidant Enzymes," *Townsend Letter for Doctors and Patients,* December 1995.

Rose, S. *The Chemistry of Life* (New York: Pelican, 1985).

GLUTATHIONE AND LIVER DETOXIFICATION

The liver is your body's largest internal organ—and for good reason. This amazing chemical factory, weighing only about three pounds, performs thousands of functions that affect your entire body. Your liver stores and produces energy, regulates hormone levels, removes old red blood cells from your system, and stores vitamins and minerals. It's also your body's primary pathway for removing metabolic wastes and poisons. To do all this, your body pumps just under a quart of blood through your liver every minute.

The metabolic processes of the liver are complex and interdependent. Consequently, your liver functions can easily be thrown out of balance. The detoxification pathways that remove wastes from your system are particularly vulnerable. If your liver isn't functioning efficiently, toxins can build up in it and elsewhere in your body. Other organs in your body that also remove wastes, such as your kidneys, are stressed as they try to cope with the overload. The result? You just don't feel right, but you don't quite know why. Your energy levels might be low; you might have digestive problems. Other common symptoms of suboptimal liver function include depression, headaches, dizziness, joint pain, sleeping disorders, dry skin, and irritability. If you're a woman, you could have severe premenstrual syndrome (PMS). The symptoms are often vague, changeable, and come and go; the usual laboratory tests for liver disease may come back normal. A traditional medical practitioner might dismiss your symptoms altogether and tell you that you are suffering from "nerves" or "stress."

A holistic practitioner knows better. The symptoms of liver detoxification problems are clear and should never be ignored. Such problems are very common today and can be readily detected with simple, usually noninvasive, and relatively inexpensive laboratory tests. Most important, health care providers can diagnose and treat low-level liver dysfunction long before it becomes liver disease or leads to cancer or diseases of the neurological, endocrine, or immune systems.

One of my patients recently provided a dramatic example of how glutathione can help. Lewis had a mysterious skin condition that dated back to 1990. His face, upper back, chest, and arms were covered with a disfiguring, horribly itchy rash. Several dermatologists had given him different diagnoses, ranging from a fungal infection to folliculitis. None of the treatments they prescribed did any good.

When Lewis finally came to me, the rash was getting worse and the itching was driving him crazy. He had to wear gloves at night to keep from scratching while he slept. Aside from the rash, Lewis was very healthy. He had plenty of energy, and his lab results were normal for a man his age.

I suspected that Lewis's liver was the problem, since unexplained, persistent rashes are sometimes an indication of poor liver detoxification function. Sure enough, the results of lab tests for a liver detoxification profile indicated that Lewis's glutathione level was very low, suggesting that his liver wasn't producing enough of this essential detoxifying enzyme. His poor liver function was letting toxins build up in his system instead of eliminating them; his body reacted with a rash.

Lewis started taking fairly high levels of GSH (glutathione), NAC (cysteine), and lipoic acid supplements, along with additional selenium and antioxidant vitamins. Within two weeks, he began to see improvements in his skin. In six months, his rash had cleared completely.

When I work with patients with liver detox problems, we have two goals. The first is to discover and treat the underlying causes of the problem. The second is to support and protect the liver. In my experience, the support your liver needs most is glutathione.

HOW YOUR LIVER WORKS

To understand why glutathione is such an essential part of the detoxification process, we once again delve into metabolic chemistry. I'll make it as simple and nontechnical as I can, because I think it's essential that you understand how your liver cleanses your system. You'll then be much more aware of how to achieve optimal health through good liver function. You'll also have a new respect for the miraculous complexity of the human body.

Your liver has two steps, or phases, for detoxifying your body. In Phase I, or oxidation (also sometimes called activation), your liver cells secrete a complex group of substances called the cytochrome P-450 enzymes. This "superfamily" of enzymes deals with the waste products of your metabolism and toxins from outside sources that are carried into the liver by the bloodstream. When the P-450 enzymes find waste molecules and toxins, they convert them to intermediate forms that are more soluble in water. This process, also called biotransformation, lets the wastes be excreted. Some go into the bile and out of the body through the colon; others are carried to the kidneys and excreted in the urine. The P-450 enzymes attack wastes in several different ways, but most are oxidized.

But doesn't oxidation create free radicals? Aren't they dangerous? It does, and they are. Phase I detoxification produces very large quantities of free radicals. In fact, it's oxidation or biotransformation in Phase I that makes many carcinogenic substances dangerous—until they're biotransformed, many are relatively harmless.

Here's where conjugation, or Phase II of liver detoxification, comes in. Conjugation is just another way of saying addition. In this phase, your liver's natural antioxidant enzymes, found in the cells and in the fluid between the cells, corral the free radicals, neutralize (conjugate) them by adding a small molecule, and escort them safely out of your body in the bile or the urine. (The

Figure 2.1

COMMON ENVIRONMENTAL TOXINS AFFECTING THE LIVER

Acetate	Food preservatives
Alcohol	Formaldehyde
Asbestos	Fungicides
Artificial food colorings	Heavy metals (aluminum, arsenic, cadmium, nickel, etc.)
Barbiturates	Herbicides
Benzene	Insecticides
Carbon tetrachloride	Organophosphorus pesticides
Cigarette smoke	Paint fumes
Dioxin	Sulfur dioxide
Exhaust fumes	Toluene

Phase II detoxification enzymes try to remove all the free radicals they find in your blood, not just those created by Phase I activity.)

Another critical function of Phase II detoxification is the removal of the biotransformed wastes created in Phase I. In addition, Phase II detoxification removes heavy metals such as cadmium, aluminum, and lead and some other toxic environmental wastes from your liver and out of your body. Toxins such as heavy metals aren't affected by Phase I oxidation, so your body depends on Phase II conjugation to remove them. And since virtually everything that you eat, drink, breathe, or even touch eventually passes through your liver, all the external toxins you inevitably encounter in daily life end up there as well (see Figure 2.1). If your Phase II conjugation activity isn't efficient, those toxins will accumulate in your fatty tissues, your brain, the coverings of your nerve cells, and elsewhere in your body—where they could later have disastrous effects on your health.

MAINTAINING A HEALTHY LIVER

Of all the antioxidant enzymes in your liver, glutathione is the most abundant and, in my view, the most important. It's the crucial enzyme for all aspects of Phase II conjugation. Therefore, it is absolutely vital to give your liver the amino acids and cofactors it needs to produce enough glutathione.

Daily Supplements for a Healthy Liver

For maintaining a healthy liver, here's what I recommend on a daily basis:

Vitamin C	500 mg
Vitamin E	400 IU
Mixed carotenes	20,000 IU
GSH	250 mg
NAC	100 mg
Lipoic acid	200 mg
Selenium	100 mcg

To simplify your antioxidant supplement regime, I recommend taking two capsules of Glutaplex daily, one in the morning and one in the evening. Manufactured by Douglas Labs, each of these capsules contains GSH, NAC, lipoic acid, selenium, riboflavin, and some other important amino acids in a base of dried parsley and spinach. Continue to take your favorite brand of vitamin C, vitamin E, and mixed carotenes.

Another important way you can support your liver and help keep it functioning well is to drink plenty of fluids. The waste products that your liver has neutralized need to be removed from your body, and by drinking a lot of fluids, the wastes are flushed out quickly through your kidneys and bowel. I recommend between 1½ and 2 quarts—six to eight 8-ounce glasses—of liquids a day. Pure water is always best, but mild herbal teas and other non-caloric or low-calorie drinks are fine too. Fruit juice is high in nutrients, but even unsweetened juice is also high in calories. The high levels of fructose can also give you diarrhea. Stick to just one or two glasses of juice a day.

In addition to adding glutathione to your daily diet and increasing your fluid intake, I recommend following a metabolic cleansing program three or four times a year. Before we get into exactly how to do that, however, you need to understand a little more about how your liver works.

UP-REGULATING YOUR LIVER

One of the most amazing things about your liver is its ability to increase its level of Phase I oxidation activity in response to an increase in your body's toxic load. This process, known as up-regulation, sounds like it ought to be beneficial. It is—but *only* if you also have a corresponding increase in your levels of Phase II detoxification. When Phase I activity increases, so does the production of free radicals and biotransformed toxins. Phase II activity has to increase as well in order to mop up all those additional reactive molecules and poisons before they cause any damage.

Liver detox problems often arise when Phase I and Phase II activity levels aren't equal. If Phase I activity is high, oxidative stress occurs: Your liver's supply of antioxidant enzymes is depleted, allowing free radicals to run wild in your body. In addition, toxins and biotransformed wastes escape into your bloodstream instead of being captured and removed from the body. The wastes can end up deposited in your fatty tissues, brain, and nerve cells. The results of poor Phase II activity can be cellular damage, low energy, an increased risk of cancer, and a wide range of neurological, endocrine, and immune system ailments.

Anyone who is exposed to an elevated toxic load is at risk for imbalance between up-regulated Phase I and Phase II activity. Look again at the environmental toxins listed in Figure 2.1. By some estimates, the average American consumes some fourteen pounds of food additives such as artificial colorings and preservatives every year. All those substances will pass through your liver. Now take the quiz in Figure 2.2. If you answered yes to two or more of the questions, you are being exposed to a high level of environmental toxins. Your liver needs extra support to keep its detoxification pathways active.

In my experience, adding glutathione or NAC to your daily diet is an excellent way to defend yourself against high levels of toxins. I also recommend taking silymarin, an herbal treatment derived

Figure 2.2

TOXIC EXPOSURE SELF-QUIZ

____ Do you smoke cigarettes or cigars or chew tobacco?

____ Are you exposed to second-hand smoke at home or work?

____ Are you frequently exposed to exhaust fumes from cars, trucks, tractors, or other internal-combustion engines?

____ Do you work in a "sealed" building with windows that can't be opened?

____ Do you live or work in a building that has new carpeting or has recently been painted?

____ Do you work with paints, solvents, or other volatile chemicals?

____ Do you use insecticides or herbicides at home or at work?

____ Do you have respiratory allergies or sinus problems?

____ Do you have joint or muscle pain or arthritis?

____ Do you have skin allergies or rashes?

____ Do you frequently take nonprescription painkillers or other medications?

____ Have you taken prescription drugs two or more times in the past year?

____ Do you have one or more alcoholic drinks per week?

____ Do you often eat grilled meat?

____ Do you eat red meat four or more times a week?

____ Are you often overtired from a normal day's activity?

____ Do you have frequent colds, minor illnesses, or infections?

If you answered yes to two or more questions, you may have a high toxic exposure that is affecting your liver.

from the milk thistle plant *(Silybum marianum)*. Extensive scientific testing has shown that silymarin functions as an antioxidant and also contains complex compounds that help protect the liver from a variety of toxins. Your liver can actually regenerate itself if it is damaged, and silymarin stimulates protein synthesis among your liver cells and helps accelerate the regeneration process. Because silymarin doesn't dissolve well in water, it is usually sold in 200 mg capsules, rather than powder form. I recommend Hepatox from

Biometrics, available at well-stocked health food stores (see Appendix B). Avoid alcohol-based silymarin extracts—you won't get enough of the active ingredients you need.

A Chinese herb called schizandra (also called wuweizi) is sometimes recommended for liver protection. In my experience, this herb is only somewhat useful, especially in comparison to silymarin, and there is less scientific evidence to back up its claims. Other herbs that are sometimes recommended for regulating the liver include goldenseal, dandelion, barberry, and yarrow. These herbs have little effect on your liver—I suggest you avoid them.

Supplements to Support Your Liver

For liver support in a toxic environment, I suggest taking these supplements on a daily basis:

Vitamin C	2,000 mg
Vitamin E	400 IU
Mixed carotenes	20,000 IU
GSH	250 mg
NAC	100 mg
Lipoic acid	200 mg
Selenium	100 mcg
Zinc	15 mg
Silymarin	200 to 600 mg

An easier alternative for your supplement regime is to take your favorite vitamin C, vitamin E, and mixed carotenes and add GSH 250 Master Glutathione Formula capsules from Douglas Labs. These capsules contain GSH, NAC, and other cofactors. I suggest taking three spread out through the day, along with silymarin.

Some people are what we call pathological detoxifiers. In response to toxins from leaky gut syndrome or from their environment, they produce high levels of Phase I cytochrome P-450 enzymes. This in turn means that they produce high levels of free radicals. Since pathological detoxifiers also have low Phase II antioxidant production, they can't quench the free radicals effectively and dangerous biotransformed wastes build up in their bodies. They often suffer from ailments such as excessive fatigue, rashes, and frequent minor illnesses such as colds. In the long run, they are at greater risk for cancer and autoimmune diseases.

To determine for certain if you are a pathological detoxifier, you may need to take some of the lab tests discussed later in this chapter. If you and your health care practitioner decide that pathological detoxification is a problem for you, you need to take some steps to boost your GSH level. I suggest 2,000 mg of vitamin C, 400 IU of vitamin E, and 20,000 IU of mixed carotenes daily. In addition, I suggest following the Phase II supplement program from the Institute of Rehabilitative Nutrition (see Appendix B). The capsules in each Phase II packet contain glutathione, NAC, sulfur, and other antioxidants designed to boost your Phase II activity. By raising your GSH level and increasing your Phase II activity, these capsules help you compensate for the higher levels of Phase I free radicals and biotransformed wastes your liver produces.

Some research suggests that drinking grapefruit juice or taking naringinen, a substance derived from grapefruit seeds, may slow Phase I activity. However, the research is still preliminary. It's too soon to tell if grapefruit juice or naringinen have any real benefit for the liver.

If your Phase I activity is too low, another set of problems can arise. If your liver isn't detoxifying well, the toxins stay in circulation in your blood instead of being excreted. They end up stored in your fatty tissues, your brain, and the cells of your central nervous system. The long-term result could be damage to your neurological, endocrine, and immune systems. Your Phase I activity could be low for several reasons. Nutrient cofactor deficiencies of riboflavin, niacin, magnesium, molybdenum, or iron are common causes; a high-fat diet is another. Other possible reasons include exposure to heavy metals, alcohol, amphetamines, or bacterial endotoxins from leaky gut syndrome or bacterial overgrowth of the small intestine.

To improve your Phase I liver function, you probably need to stimulate your cytochrome P-450 activity. Good ways to do this are to avoid recreational drugs and alcohol, correct any nutritional deficiencies with supplements and a better diet, avoid refined sugar, and eat a nutritious, low-fat diet with plenty of high-quality protein. If bacterial endotoxins are causing the problem, consider the treatment options we'll discuss later in this chapter. In addition, I suggest following the Phase I supplement program from the Institute of Rehabilitative Nutrition. The capsules in each Phase I packet contain NAC, quercetin (a powerful antioxidant), and other substances designed to boost your Phase I activity. In addition, continue to take 2,000 mg of vitamin C, 400 IU of vitamin E, and 20,000 IU of mixed carotenes. You should also add 600 to 800 mg of silymarin as a liver protectant.

If Phase II (conjugation) activity is low, oxidative stress occurs. If both Phase I and Phase II activity levels are low, you have what is commonly called a sluggish liver. This is not good—it means you have lots of circulating toxins and free radicals along with insufficient antioxidants. Serious illness may be much more likely to develop. If your liver is very sluggish, your glutathione level may be dangerously low. And when you don't have enough glutathione, your liver isn't capturing all those free radicals, it's not removing biotransformed wastes from Phase I, and it's not removing environmental toxins. Although dietary changes and supplements can do a lot to improve your liver function, a sluggish liver needs more help.

METABOLIC CLEANSING

A metabolic cleansing program can be extremely helpful for patients with a sluggish liver—and even for those whose liver function is only slightly out of balance. Many of my patients have experienced dramatic improvements in their health after following the complete program. Metabolic cleansing helps to remove accumulated toxins from your body and restore a healthy balance of liver functions.

Should you try a metabolic cleansing program? Complete the testing scale in Figure 2.3 and add up your score. This scale was designed by Dr. Jeffrey Bland, a leading researcher in the area of metabolic clearing therapy. In my experience, it gives a very accurate rating of liver function. If your score is 25 or greater, your liver function is probably below par and you could almost certainly benefit from following the program.

Figure 2.3

LIVER FUNCTION TESTING SCALE

Rate each of your symptoms based upon your typical health profile
for the past thirty days.

0   Never or almost never have the symptom
1   Occasionally have the symptom; effect is not severe
2   Occasionally have the symptom; effect is severe
3   Frequently have the symptom; effect is not severe
4   Frequently have the symptom; effect is severe

Bowel:

diarrhea
constipation
gas
undigested foods in stool
irritable bowel syndrome

Digestive Tract:

nausea or vomiting
bloated feeling
belching/gas
heartburn

Ears:

itchiness
earaches or ear infections
drainage from ear
ringing in ears or hearing loss

Emotions:

mood swings
anxiety, fear, or nervousness
anger, irritability, or aggressiveness
depression

Energy/Activity Levels:

fatigue or sluggishness
apathy or lethargy
hyperactivity
restlessness

LIVER FUNCTION TESTING SCALE (cont.)

Eyes:

watery or itchy eyes
swollen, reddened, or sticky eyelids
bags or dark circles under the eyes
blurred or tunnel vision (aside from near- or farsightedness)

Head:

headaches
faintness
dizziness
insomnia

Heart:

irregular or skipped heartbeats
rapid or pounding heartbeats
chest pain

Joints/Muscles:

pain or aches in joints
arthritis
stiffness or limitation of movement
pain or aches in muscles
feeling of weakness or tiredness

Lungs:

chest congestion
asthma or bronchitis
shortness of breath
difficulty breathing

Mind:

poor memory
confusion or poor comprehension
poor concentration
poor physical coordination
difficulty in making decisions
stuttering or stammering
slurred speech
learning disabilities

LIVER FUNCTION TESTING SCALE (cont.)

Mouth/Throat:

> chronic coughing
> gagging or frequent need to clear throat
> sore throat, hoarseness, loss of voice
> swollen or discolored tongue, gums, lips
> canker sores

Nose:

> stuffy nose
> sinus problems
> hay fever
> sneezing attacks
> excessive mucus formation

Skin:

> acne
> hives, rashes, or dry skin
> hair loss
> flushing or hot flashes
> excessive sweating

Weight:

> binge eating or drinking
> food cravings
> excessive weight
> compulsive eating
> water retention
> underweight

Other:

> frequent illness
> frequent or urgent urination
> genital itch or discharge

Total: _____

Source: Jeffrey Bland, Ph.D., HealthComm, Gig Harbor, WA.

Please read this carefully: Metabolic cleansing programs are not designed for weight loss. They should not be used by children, pregnant women, diabetics, or anyone with hypoglycemia, kidney disease, liver disease, or any other serious medical problem. If you take medication for high blood pressure, a heart condition, or any other chronic ailment, discuss metabolic cleansing with your physician before you try it.

If you feel you could benefit from a liver cleansing program, don't do it on your own—work with a nutritionally oriented health care practitioner. The products I recommend below for your program are available only through such health care providers and must be used under their supervision. Accept no substitutes.

Two manufacturers make complete supplement mixtures that provide the liver support you need as you go through the program: Ultra Clear from HealthComm and Bio Detox Powder from the Institute of Rehabilitative Nutrition. These products provide the vitamins, minerals, amino acids, and cofactors you need to boost your antioxidant levels, increase your glutathione level, and help your liver regain its normal function. The supplements are in a hypoallergenic rice base; they contain no corn, wheat, soy, yeast, egg, dairy, or animal products. They're easy to use: Just mix two scoops (60 g) into eight ounces of cold filtered or bottled water or diluted fruit juice, and drink. One serving has about 150 calories. Additional supplements such as psyllium fiber or lactobacillus (see below) can be added if desired. I also recommend taking three capsules a day of GSH 250 Master Glutathione Formula.

The full liver cleansing program has three steps, which should always be done under supervision. In severe cases, you may have to follow all three steps and take up to twenty-one days to complete the program. In most cases, however, you can complete the program in just a week or two. It's easy to do. Simply follow the dietary guidelines outlined below and take the detoxification mixtures as recommended.

PRIMARY CLEANSING. If your liver problems are severe (a score of more than 100 on the testing scale), begin your program with primary cleansing. This requires a liquid diet and detoxification supplements for one to five days. Prepare a few days in advance by eating a light, low-fat diet with lots of fresh fruits and vegetables; avoid caffeine. Begin the program in the morning with one detox drink; have three to four more at intervals throughout the day. In between, drink plenty of filtered or bottled water, herbal tea, or diluted fruit juice. It's extremely important to keep well hydrated during every phase of the program, but it's especially important in this phase. Including the detox drinks, be sure to take in sixty-four to eighty ounces of liquid a day. In addition, take three GSH 250 Master Glutathione Formula capsules, spread out over the day.

You may well be wondering why you need the detox drinks. Couldn't you just go on a liquid fast without them? At one time, many alternative health care practitioners did recommend liquid fasting for several days as a way to detoxify the liver. By not eating and drinking only pure water or diluted fruit juices, you supposedly reduced the strain on your liver and allowed it to "concentrate" on detoxification. In fact, a solid body of research proves that in many cases, fasting doesn't help and can even actually impair the detoxification process. When you fast, your cytochrome P-450 activity drops sharply, allowing toxins to pass through your liver unchanged. Your normal supplies of antioxidants are quickly depleted when you fast, putting you into a state of oxidative stress. Instead of helping, then, fasting just lets toxins and free radicals roam your body, doing damage all along the way.

The detoxification supplements taken as part of your liquid diet in the primary cleansing provides the additional glutathione, other antioxidants, and cofactors you need to avoid oxidative stress during this stage. This way you really do get the benefit of fasting—a reduced load on your liver—while avoiding the potential damage.

How long should you stay in the primary cleansing stage? That depends on how you feel. In general, my patients stay on the liquid diet for only two to three days. At that point, they are already feeling more energetic and are ready for the next step. Do not continue primary cleansing for more than five days. Remember, *this is not a weight-loss diet.* Talk to your health care provider about how long you should stay in the primary phase. If you have un-

Figure 2.4

SECONDARY CLEANSING FOODS

Choose six servings a day from the following foods. A serving is one piece (one apple, for example) or approximately one cup (8 ounces) except where otherwise noted.

apple	olive oil (1 tablespoon)
banana	papaya
broccoli	peach
carrot	pear
cauliflower	peas
celery	sweet potato/yam
cucumber	sunflower oil (1 tablespoon)
flaxseed oil (1 tablespoon)	white rice
green beans	winter squash
kiwi fruit	zucchini
melon	

Note: All vegetables should be lightly steamed. Avoid salt, sugar, spices, and condiments.

pleasant symptoms such as headaches or joint pain, discuss the problem with your health care provider. You may need to move on to the next phase ahead of schedule.

SECONDARY CLEANSING. If your liver function is not severely out of balance, you can probably skip the primary cleansing and go directly to the secondary stage. During this stage, have three detox drinks a day. Keep well hydrated by drinking a daily total (including the detox drinks) of sixty-four to eighty ounces of filtered or bottled water or diluted fruit juice. In addition, choose at least six servings a day from the foods in Figure 2.4. A serving is one piece (one apple, for example) or approximately one cup (eight ounces), but there is *no limit* to how much you can eat of these foods. A metabolic cleansing program is not a weight-loss diet—you should not feel hungry. Take three GSH 250 Master

Glutathione Formula capsules, spread out over the day. Continue with this stage of the program for up to seven days.

FINAL CLEANSING. If you have only mild liver imbalance, you can probably proceed directly to this stage of metabolic cleansing. In fact, I recommend this stage for everyone at least once a year. During this stage, have one to two detox drinks a day and take three GSH 250 Master Glutathione Formula capsules, spread out over the day. Keep well hydrated by drinking a daily total (including the detox drinks) of sixty-four to eighty ounces of filtered or bottled water or diluted fruit juice. You can now add mild herbal teas as long as they are not citrus-based; continue to avoid caffeine. Choose at least six servings a day from the foods in Figure 2.5. In this stage, the protein and grain servings are up to you—eat as much as you want. Continue with this stage of the program for seven to ten days.

WHAT TO EXPECT. As your liver removes accumulated toxins over the course of the metabolic cleansing program, you may feel some minor discomfort during the first few days. Common symptoms might include fatigue, poor stamina, light-headedness, headaches, insomnia, gas and bloating, constipation, diarrhea, skin irritation, muscle aches, or painful joints. These symptoms are usually minor and pass quickly; in fact, many of my patients don't experience any discomfort at all. Discuss your symptoms with your health care practitioner if they occur. If they are severe, you may need to move on to the next step in the program or go off it completely. If they are not too bothersome, however, I urge you to continue with the program. Once the symptoms pass, you may well experience a level of well-being you haven't felt in years.

One of my patients provides a very good example of how helpful metabolic cleansing can be. Harley had been complaining about gas, bloating, burping, and diarrhea for two years. His doctor had done a full gastrointestinal workup, but everything appeared normal. Harley had no intestinal parasites or other stool abnormalities. Antacids and other medications didn't seem to help the symptoms. When Harley started having joint and muscle pain in addition to his gastric distress, he came to see me.

Figure 2.5

FINAL CLEANSING FOODS

Choose six servings a day from the following foods. A serving is one piece (one apple, for example) or approximately one cup (8 ounces) except where otherwise noted.

Protein:

cod	chicken (skinless)
halibut	turkey (skinless)
sole	eggs
tuna	kidney, soy, lima beans
lamb	

Grains:

rice bread	cream of rice
rice cakes	rice flour pancakes
brown rice	millet

Vegetables:

alfalfa sprouts	kale
artichoke	kohlrabi
asparagus	leeks
beets	lettuce
bell peppers	okra
(sweet)	onion
bok choy	parsley
broccoli	parsnip
brussels sprouts	potato
cauliflower	radish
chard	spinach
daikon radish	sweet potato/yam
eggplant	taro
endive	turnip
escarole	water chestnuts
green beans	
jicama	

FINAL CLEANSING FOODS (cont.)

Fruits:

apple	papaya
apricot	plums
avocado	raspberries
blueberries	strawberries
melons	

Spices:

bay leaf	ginger
caraway seeds	mace
chives	marjoram
cinnamon	mint
curry	nutmeg
dill	poppy seeds
dry mustard	savory
garlic	tarragon

Note: All meat and fish should be baked. Eggs should be boiled or poached. All vegetables should be lightly steamed.

His nutritional evaluation revealed that his gastrointestinal tract was indeed normal. A lab test called the functional liver detoxification profile, however, revealed that Harley had impaired Phase I and Phase II detoxification pathways. Harley was a slow detoxifier.

This pattern suggested that Harley had a metabolic difficulty in processing and removing toxins from his body. Normal variations in liver enzymes may have been the root cause of the problem, but a shortage of nutritional cofactors such as selenium made it worse.

Harley went on a program to up-regulate his liver function. First, he completed the metabolic cleansing program. Next, he began taking daily supplements of NAC, GSH, lipoic acid, and selenium, in addition to antioxidant vitamins. Within sixty days, Harley had no symptoms. As long as he continues to take his antioxidant supplements regularly, Harley continues to feel great.

CAUSES OF DETOX PROBLEMS

What makes your liver functions get out of balance? There are a lot of answers to that crucial question.

ENVIRONMENTAL TOXINS. I think environmental toxins are the primary culprit behind many cases of poor liver function. It is virtually impossible to go through a day, much less a lifetime, in modern society without encountering numerous environmental toxins. As you've already seen in Figure 2.1, we are constantly exposed to these poisons in the course of daily life.

Your liver isn't really designed to cope with such a large toxic load. Even if you make a real effort to avoid such toxins, your liver can still easily become overburdened. In an effort to deal with the load, your liver will up-regulate itself and start producing a lot more cytochrome P-450 enzymes. That means you're also producing a lot more damaging free radicals and biotransformed wastes. And if the toxic load is large enough, even up-regulated Phase I detoxification won't be enough, allowing many toxins to leave the liver untouched.

A second major cause of liver detoxification imbalance could come from within your own body—your intestines, to be precise. Your body absorbs both toxins and nutrients through the lining of the small intestine. Of course, as we've already discussed, any toxins you consume (foods treated with pesticides or herbicides, for example) will be absorbed into your blood and eventually end up in your liver for detoxification.

LEAKY GUT SYNDROME. Your small intestine has two contradictory roles. The nutrients your body needs are absorbed through the mucus membrane of the intestinal lining. But that same lining is also a barrier to keep out toxins, bacteria, large molecules, and particles of undigested food. Ordinarily, your intestinal lining is very "tight." Only small molecules of digested nutrients, such as amino acids and peptides, should be able to pass through the membrane of the gut and be absorbed into your bloodstream. If your small intestine has been damaged in some way (by illness or certain drugs, for example), or if you have food allergies, you could develop leaky gut syndrome. In that case, your small intestine becomes more permeable and allows toxins and larger molecules of incompletely digested food to enter your bloodstream. Your body often reacts to the large molecules (also called endotoxins, or toxins from within) as if they were allergy triggers. If you have leaky gut syndrome, you might develop symptoms of allergic reaction, including inflammation, rashes, diarrhea, joint pain, or even asthma. In more serious cases, you could develop an illness such as Crohn's disease, rheumatoid arthritis, psoriasis, or some other chronic, debilitating condition. Some illnesses, or even the treatments for them, can cause leaky gut syndrome: Examples include inflammatory bowel disease, HIV infection or AIDS, pancreatic insufficiency, celiac disease, food allergies, and parasite infestation. If you have been treated with nonsteroidal anti-inflammatory drugs (NSAIDs), steroid drugs, or antibiotics, or if you have been frequently exposed to environmental toxins or alcohol, you could have leaky gut syndrome. And as we'll discuss below, toxins from bacterial overgrowth of the small intestine can also lead to leaky gut syndrome.

Leaky gut syndrome puts a lot of strain on your liver. First, more toxins are entering your body, which will have to be detoxified when they reach your liver. Your cytochrome P-450 enzymes will be activated, which in turn means that you will produce more free radicals and need more antioxidants. In addition, your body is producing antibody substances in reaction to the large molecules that have entered your bloodstream. Since your liver is the organ that copes with the waste products of antibodies, your Phase I and Phase II activity levels jump. In addition, antibody reactions increase your overall need for antioxidants, especially glutathione.

Health practitioners are realizing that leaky gut syndrome is at the root of a lot of chronic poor health and that it is fairly common. For many of my patients, the cause is prescription or over-the-counter nonsteroidal anti-inflammatory drugs (NSAIDs) such as aspirin and ibuprofen (Advil). Ironically, NSAIDs are commonly prescribed to treat arthritis, but in some cases they could actually be making the problem worse. Other possible causes of increased

Figure 2.6

LEAKY GUT SYNDROME SYMPTOMS

bloating	fatigue
gas	food allergies
cramps	skin rashes
diarrhea	headaches
constipation	

intestinal permeability include intestinal infections, yeast infections, parasites such as giardiasis, alcoholism, aging, food allergies, and environmental toxins.

The underlying cause of leaky gut syndrome is a shortage of the amino acid glutamine, which is actually a form of glutamic acid, one of the three amino acids that make up glutathione. Glutamine is the primary nutrient for the cells that line your small intestine. In fact, it is so important to intestinal health that some researchers, citing the importance of glutamine in the treatment of critically ill patients, suggest it should be called a "conditionally essential" amino acid. If you don't get enough glutamine for any of the reasons discussed above, the mucus lining of your small intestine thins out and becomes permeable.

A low glutamine level in the small intestine means a low glutathione level everywhere in your body, since one of the essential building blocks of glutathione is in very short supply. This can lead to a nasty cycle: Low glutamine leads to leaky gut syndrome, which places an increased load on the liver and creates a higher demand for glutathione. An increased need for glutathione robs the small intestine of glutamine, which makes the leaky gut syndrome worse. And so it goes, onward and downward to ever-worsening health.

If you think that a leaky gut is causing some of your health problems, look at the symptoms listed in Figure 2.6. If you have had more than two of the symptoms over the past four to six weeks, leaky gut syndrome could be the cause. You'll need to discuss the problem with your health care practitioner to discover the causes and design a treatment program. One major cause of leaky gut is bacterial overgrowth of the small intestine. This is such an important cause that we'll discuss it separately below. Since food allergies are another significant cause of leaky gut syndrome, you might want to investigate these carefully. Many of my patients benefit from reducing the amounts of sugar and carbohydrates in their diets and eating more fresh vegetables and protein. In addition to any lifestyle and dietary changes you decide to make, I recommend taking three GSH 250 Master Glutathione Formula capsules a day, along with 5,000 mg of supplemental glutamine. Pure L-Glutamine—Intestinal Permeability Powder, manufactured by Primary Nutraceuticals—is a reliable brand.

Leaky gut syndrome doesn't go away quickly. It could take a while to determine the causes and longer to repair the damage. Once you get on the right track with diet and supplements, however, your leaky gut problems could be reduced or gone in one to three months.

BACTERIAL OVERGROWTH OF THE SMALL INTESTINE. The real work of digestion in your small intestine is done by billions of "friendly" bacteria. In fact, you have many more bacteria in your gut than you have cells in your body! In the healthy gut, literally hundreds of different kinds of bacteria work together to break down your food into its components. The delicate balance of bacteria can be disrupted, however, by illness, injury, drugs, insufficient stomach acid (hypochlorhydria), or poor diet—alone or in combination. The altered intestinal environment can favor the growth of one type of bacteria over all the others; it can also let yeast organisms such as *Candida albicans* get out of control. When undesirable gut organisms badly outnumber friendly organisms, you can't digest your food or absorb nutrients properly. You suffer digestive problems such as gas or diarrhea, and your vitamin levels and overall nutritional status drop. In addition, bacterial intestinal overgrowth, known as dysbiosis, can lead to leaky gut syndrome. Over the long term, dysbiosis can lead to other severe problems such as osteoporosis. And finally, if you don't get enough of the amino acid building blocks and cofactors, you can't produce enough P-450 enzymes or antioxidant enzymes. The end result of this major burden on your liver is oxidative stress as damaging toxins build up in your body.

As with leaky gut syndrome, bacterial overgrowth of the small intestine is an ailment more widespread than many health practitioners realize. Compare your symptoms to those in Figure 2.7. If you have had more than two of the symptoms over the past six weeks, bacterial overgrowth could be the problem. If you have any of the chronic conditions listed, bacterial overgrowth could be a contributing factor. In either case, you may need simple, noninvasive lab tests to verify bacterial overgrowth. (Lab tests are discussed later in this chapter.)

One of my patients, Lisa, is a good example of how intestinal overgrowth can cause myriad health problems. At age thirty-four, Lisa felt old. She was losing weight, her joints ached, and her muscles were weak. Her doctor found that she was slightly anemic, but not enough to be causing her symptoms. He referred her to an internist, who did a thorough examination and a number of blood tests but failed to find anything specifically wrong. A nurse at the internist's office referred Lisa to me.

We began by doing a clinical assessment of Lisa's nutritional status. She scored high on the dysbiosis scale, indicating a gastrointestinal problem. A comprehensive digestive stool analysis (CDSA) revealed that she had an imbalance in her intestinal flora. In addition, her intestinal acidity was somewhat high. Her weight loss was occurring because she was no longer absorbing nutrients properly through the intestinal walls.

Lisa was put on a program of intestinal reinoculation to restore the proper balance of her intestinal bacteria. She started taking a beneficial bacteria formula every day, along with supplements of fructooligosaccharides (FOS) to help the bacteria take hold. She also took a daily dose of glutamine to help normalize her intestinal acidity. Within two months, Lisa was feeling a lot better. She regained the weight she had lost, her joints stopped aching, and her energy levels returned to normal. Lab tests showed that her intestinal overgrowth had been overcome. Lisa has been fine ever since.

If bacterial overgrowth is causing problems for you, I suggest you work with a holistic health care practitioner to identify the causes and modify your diet if needed. In mild cases, avoiding refined sugar in any form, eating a diet low in fat and high in

Figure 2.7

BACTERIAL OVERGROWTH SYMPTOMS

Within One to Two Hours of Eating:
gas
bloating (especially in lower abdomen)
diarrhea
cramps

Over a Period of Several Days or More:
weight loss
fatigue
depression
difficulty concentrating
anemia
increased susceptibility to infection
PMS
vaginal yeast infection
respiratory allergy symptoms (sinusitis, rhinitis, bronchitis)
skin allergy symptoms (eczema, dermatitis)

Source: Great Smokies Diagnostic Laboratory, 1996.

complex carbohydrates, and adding fermented dairy products such as live-culture yogurt can help. In more serious cases, however, you will need to take further steps to get your intestinal microbes back into a healthy balance.

I find that many of my patients benefit from a two-step approach that adds beneficial bacteria and also provides nutrients that help bacteria become reestablished in your gut. To add beneficial bacteria, I recommend taking high-quality acidophilus, bulgaricus, bifidobacteria, or lactobacillus in powder form. A confusing number of brands are available at your favorite health food store. As a guide, select only products that:

- Contain only one type of bacteria per container
- Contain the DDS-1 acidophilus strain or the Malyoth bifidobacteria strain
- Were cultured in a milk-based medium

- Have been ultra-filtered, not centrifuged
- Are certified to contain at least one billion (yes, *billion*) active bacteria per gram
- Have been kept refrigerated
- Do not contain the bacteria *L. casei* or *Streptococcus faecium*

My patients get the maximum benefit from friendly bacteria supplements if they combine acidophilus with bifidobacteria. For maintaining good intestinal balance, I suggest a daily treatment of 1 g (about half a level teaspoon) of acidophilus with 250 mg (about an eighth of a teaspoon) of bifidobacteria. Mix the powders in eight ounces of pure, chlorine-free cold water and drink it on an empty stomach about ten to fifteen minutes before eating. Taking the bacteria on an empty stomach seems to help them survive the powerful acids of your stomach and the first part of your small intestine and arrive safely in the lower portion of your small intestine. If you prefer a simpler way to take your beneficial bacteria supplements, I recommend Prodophilus Complex from Primary Nutraceuticals. Mix a quarter teaspoon from each of the three bottles into three ounces of pure, chlorine-free cold water. Drink it daily, ten minutes before eating breakfast.

If you have a serious overgrowth or if your overgrowth is affecting your liver function, you may need to take between 5 and 10 g of acidophilus and up to 6 g of bifidobacteria daily, spread out between meals during the day. In addition, you may wish to add supplements of *Lactobacillus bulgarica*. This bacteria is most effective when 3 to 6 g are taken with meals.

Once the bacteria arrive in the proper part of your small intestine, they need some nutritional help to regain their foothold. I strongly recommend taking supplements of fructooligosaccharides (FOS) along with your beneficial bacteria supplements. FOS fuel the growth of beneficial bacteria in the gut. Because they are not broken down and digested by your body, they are completely available only to the bacteria. Low levels of FOS are found naturally in some foods such as honey and garlic; artichoke flour contains higher levels of FOS. To get the real benefits of FOS I suggest you use it in the form of 95 percent pure powder or syrup. FOS products are mildly sweet and can be used as a sweetener in beverages or on food (try it on yogurt), although they don't work very well as a sugar substitute in cooking. I recommend the powder or tablets from Allergy Research Group (see Appendix B). I suggest starting with 1 g (about a quarter teaspoon) a day and gradually increasing the dose to 3 to 4 g daily. Large amounts of FOS can cause mild diarrhea in some people. If this occurs, reduce your daily dose until you return to normal.

Finally, I suggest adding glutamine supplements. Glutamine is vital for nourishing the cells that line your small intestine, but bacterial overgrowth can keep glutamine from reaching the lining. As we've already seen, this can lead to leaky gut syndrome.

Dysbiosis can take several weeks or even months to clear up. It's important to keep taking your supplements every day even if you don't notice an immediate improvement. If you keep with it, you'll probably start to notice a slight improvement after the first couple of weeks and a more significant improvement in six weeks. Once you're back to normal, stop taking the supplements.

STRESS. Now that you have a good understanding of how your liver's detoxification system works, you can appreciate how illness or poor nutrition can throw it badly off balance. What you may not realize is how harmful stress can be to your liver. When you are under stress, you create high levels of hormones, such as adrenaline, estrogen, and testosterone, that are later broken down in the liver by a process called sulfation. This process biotransforms the steroid hormones, prostaglandins, and phenol-based compounds your body naturally produces (and also some of the drugs you take) into less harmful forms that can then be excreted safely.

Sulfation is critically dependent on an adequate level of glutathione. If you don't have enough glutathione, you don't have enough of the sulfur-containing amino acid cysteine, and your liver can't perform its sulfation function properly. If you're under a lot of stress, not only are you producing lots of extra hormones from amino acid building blocks, you're also probably not eating properly. The combination leads to extra stress on your liver as it uses up glutathione to dispose of the excess hormones through sulfation. At the same time, your liver also needs glutathione for

Phase II activity. If your glutathione level is too low, the sulfation process will be slow, your liver won't be able to cope with all the free radicals and biotransformed wastes created by Phase I activity, and it won't be removing toxic wastes and heavy metals efficiently. Stress can easily lead to a downward spiral of ever-worsening health as your body builds up metabolic wastes, excess toxins, and free radicals.

If your liver detox functions are overloaded from stress, you need to get your complete B-complex vitamins (especially thiamine and riboflavin), along with the supplements I recommended for a high load of liver toxins on page 25. In addition, I recommend two capsules daily of Mitochondrial Resuscitate, a stress formula developed by Dr. Jeffrey Bland of HealthComm. Take them as long as the stressful situation lasts.

OTHER LIVER PROBLEMS. Another significant reason for poor liver function is liver damage from alcohol and drug abuse, and illnesses such as cirrhosis or hepatitis. If you have liver damage of this sort, glutathione supplements, silymarin, vitamin supplements, and additional nutritional support are almost certainly called for. Discuss your condition with your health care provider, however, before adding any herbs or supplements to your diet.

LAB TESTS FOR LIVER FUNCTION

You can get your liver functions back into balance more easily if you know exactly why they are out of balance to begin with, and to what degree. Several different laboratory tests can provide a very accurate picture of your intestinal status and liver function.

If you and your health care practitioner suspect leaky gut syndrome, I suggest you take an intestinal permeability test to confirm the diagnosis and the severity of the problem. You can do this with a simple, noninvasive lab test at home using a kit your health care practitioner can order from a testing lab (see Appendix A).

The first step for taking the intestinal permeability test is to collect a urine specimen in the container provided with the kit at any convenient time the day before you do the test. That night, do not eat or drink anything after 11 P.M. The next morning, mix the challenge drink according to the instructions provided and drink it on an empty stomach. Carry on with your normal activities for the next six hours, and then collect another urine specimen. Package the specimens according to the kit's directions and send them off to the lab. Your health care practitioner will receive the results in forty-eight hours.

The challenge drink in the intestinal permeability test contains two different sugars that are water-soluble but are not actually metabolized by your body: mannitol and lactulose. Most people readily absorb all or most of the mannitol through the small intestine but absorb very little of the lactulose. Because mannitol is absorbed but not metabolized, it will later be excreted unchanged in your urine. Because very little lactulose is absorbed through the gut, almost none will be excreted in the urine—it will simply pass through the gut unabsorbed and be excreted in the feces instead.

If you have a leaky gut, however, you will absorb more lactulose than normal through the small intestine, which means that you will excrete more of it than normal in your urine. If you have an increased lactulose recovery rate—more lactulose than normal in your urine—you may have increased intestinal permeability. If you have an increased mannitol recovery rate—if you excrete it faster than normal in your urine—this too could indicate increased intestinal permeability. If you have a decreased mannitol recovery rate—if you have less mannitol than normal in your urine—it could be because you haven't absorbed as much as normal. This could indicate malabsorption, perhaps because of bacterial overgrowth of the small intestine.

Another easy lab test you can perform at home, the functional liver detoxification profile, can help confirm whether you have a liver detox problem and determine how severe it is. If you think you have a liver problem, I suggest you speak to your holistic health practitioner about doing this test. He or she can arrange for a test kit and can later explain the results to you (see Appendix A for labs that do this testing).

The functional liver detoxification profile measures how quickly you clear a small, premeasured amount of caffeine from your system. It also measures how quickly you clear a small, premeasured amount of aspirin combined with acetaminophen.

To do the test, you must fast overnight—that means eating your normal evening meal and having nothing to eat or drink after 11 P.M. that night. The next morning, drink the challenge mixture first thing, before you eat or drink anything else. For the rest of the day, it is very important to avoid caffeine in any form—coffee, tea, caffeinated soft drinks, maté, chocolate, and so on. Also avoid any over-the-counter drugs containing aspirin, acetaminophen (Tylenol), or caffeine in them (many cold or allergy medications, for example). Read the labels carefully.

One hour after taking the drink, collect a urine specimen in the container provided with the test kit. Wait another hour, then collect a saliva specimen in the container provided with the test kit. Collect a second saliva specimen six hours later. Package the specimens as directed by the test kit and send them to the lab. The results will come back to your health care provider within forty-eight hours.

What are you actually testing? The urine sample measures how quickly the aspirin and acetaminophen clear your system. Aspirin is metabolized in the liver by both Phase I (activation) and Phase II (conjugation). Acetaminophen is metabolized by Phase II. The urine sample measures the amounts of biotransformed wastes excreted by the metabolism of the aspirin and acetaminophen doses. The measurements provide a detailed picture of Phase I cytochrome P-450 activity and Phase II sulfation and glutathione conjugation activity. If, for example, you excrete low levels of the waste product acetaminophen mercapturate, you are probably not producing enough glutathione.

The urine sample also measures the ratio of sulfate to creatinine in your urine. Creatinine is an enzyme waste product that is excreted through the kidneys. If the ratio is high, your level of glutathione is probably adequate. If the ratio is low, it suggests that you don't have enough glutathione or sulfate in your system.

The saliva samples measure how quickly caffeine clears your system. Caffeine is completely absorbed by the small intestine and is metabolized in the liver by the P-450 enzymes in the Phase I process. The rate at which you metabolize the caffeine indicates the amount of P-450 enzymes you have and the level of Phase I activity.

In general, if your Phase I levels are high—if you clear the caffeine very rapidly—your liver has been up-regulated, probably from exposure to environmental toxins. Your health care practitioner will probably say that your Phase I system is induced. If your Phase I levels are low—if you clear the caffeine slowly—your liver needs to be up-regulated.

If you'd like to take your liver function testing a little further, I suggest you add an oxidative stress panel to your lab tests. This panel consists of two blood tests, so you will have to have blood drawn by a health care practitioner. The blood is tested for its level of whole blood reduced glutathione and for lipid peroxides. The level of glutathione in the blood is a helpful indicator of your overall level of oxidative stress. The more glutathione in your blood, the more your oxidative stress from free radicals—because your body is producing lots of glutathione to fight the free radicals. The lipid peroxide test is a way to measure your free radical levels. Lipid peroxidation occurs when free radicals damage the fatty membranes of your cells. If you have a lot of lipid peroxidation byproducts in your blood, you have a lot of free radicals zooming around, doing damage to your cells. It also means that you are low on the antioxidants that should be mopping up the free radicals.

If you suspect that you have leaky gut syndrome or bacterial overgrowth of the small intestine, I suggest you speak to your health care practitioner about tests to confirm the problem. For the most informative results, I recommend a comprehensive digestive stool analysis (CDSA). This easy, noninvasive test reveals the presence of leaky gut syndrome and malabsorption markers. It also analyzes the bacteria present in the gut. If there is an overgrowth, it tells you exactly which bacteria are too abundant. Your health care practitioner can provide you with a test kit. All you have to do is collect two or three stool samples in the privacy of your home, prepare them as explained in the kit, and send off the samples in the packaging provided. For a simpler but somewhat less informative test, consider a breath analysis. Your health care practitioner can provide you with the test kit. This noninvasive test requires you to eat a low-fiber diet for one day, have a light evening meal, and then fast overnight. The next morning, collect a breath sample in a special tube provided in the kit. You then swallow a small, premeasured amount of lactulose or glucose and collect more breath samples every fifteen minutes for two hours, for a total of eight samples. The samples are sent to the lab and analyzed for the presence of hydrogen and methane gas over time. If your hydrogen and methane levels are high, bacterial overgrowth is probably present. The higher the levels, the more severe the problem.

I have treated many patients suffering from poor liver function and the related problems of leaky gut syndrome and bacterial overgrowth of the small intestine. Most of them start to feel much better within one to two months of modifying their diet and taking daily supplements of GSH and other nutrients. Some patients, however, don't improve much at all with this treatment. These puzzling patients have a more subtle problem that requires a somewhat different approach: The environment around them is making them sick.

REFERENCES

Anderson, K. E., and A. Kappas. "Dietary Regulation of Cytochrome P-450," *Annual Review of Nutrition,* vol. 11, pp. 141–46, 1991.

Benuck, M., et al. "Effect of Food Deprivation on Glutathione and Amino Acid Levels in Brain and Liver of Young and Aged Rats," *Brain Research,* vol. 678, no. 1–2, pp. 259–64, April 1995.

Bland, Jeffrey S. "Food and Nutrient Effects on Detoxification," *Townsend Letter for Doctors and Patients,* December 1995.

Bland, Jeffrey S., and Alexander Barlley. "Nutritional Up-Regulation of Hepatic Detoxification Enzymes," *Journal of Applied Nutrition,* vol. 44, no. 3–4, pp. 1–15, 1992.

Chasseaud, L. F. "The Role of Glutathione and Glutathione S-transferases in the Metabolism of Chemical Carcinogens and Other Electrophilic Agents," *Advances in Cancer Research,* vol. 29, pp. 176–244, 1975.

Crook, William G., M.D. *The Yeast Connection* (New York: Vintage Books, 1986).

Godin, D. V., and S. A. Wohaieb. "Nutritional Deficiency, Starvation, and Tissue Antioxidant Status," *Free Radical Biology and Medicine,* vol. 5, pp. 165–76, 1988.

Grattagliano, I., et al. "Effect of Oral Glutathione Monoethyl Ester and Glutathione on Circulating and Hepatic Sulfhydrils in the Rat," *Pharmacological Toxicology,* vol. 75, no. 6, pp. 343–47, 1994.

Gregus, Z., et al. "Effect of Lipoic Acid on Biliary Excretion of Glutathione and Metals," *Toxicology and Applied Pharmacology,* vol. 114, no. 1, pp. 88–96, May 1992.

Guengerich, F. P. "Effects of Nutritive Factors on Metabolic Processes Involving Bioactivation and Detoxication of Chemicals," *Annual Review of Nutrition,* vol. 4, pp. 207–31, 1984.

Hunnisett, Adrian, et al. "Gut Fermentation Syndrome . . . ," *Journal of Nutritional Medicine,* vol. 1, pp. 33–38, 1990.

Kaplowitz, N. "The Importance and Regulation of Hepatic Glutathione," *Yale Journal of Biology and Medicine,* vol. 54, pp. 497–502, 1981.

Kerr, Kevin G. "The Gastrointestinal Microflora: Friends or Foe?" *Journal of Nutritional Medicine,* vol. 3, pp. 39–44, 1991.

Lacey, J. M., and D. W. Wilmore. "Is Glutamine a Conditionally Essential Amino Acid?" *Nutrition Review,* vol. 48, no. 8, pp. 297–307, 1990.

Madara, J. L. "Pathobiology of the Intestinal Epithelial Barrier," *American Journal of Pathology,* vol. 137, no. 6, pp. 1273–81, 1990.

Meister, A. "Selective Modification of Glutathione Metabolism," *Science,* vol. 220, pp. 43–47, 1983.

Metcalfe, D. D. "The Nature and Mechanisms of Food Allergies and Related Diseases," *Food Technology,* vol. 46, pp. 136–39, 1992.

Reed, D. J., and P. W. Beatty. "Biosynthesis and Regulation of Glutathione: Toxicological Implications," *Review of Biochemistry and Toxicology,* vol. 2, pp. 213–41, 1980.

Salmi, H. A., and S. Sarna. "Effect of Silymarin on Chemical, Functions, and Morphological Alterations of the Liver," *Scandinavian Journal of Gastroenterology,* vol. 17, pp. 517–21, 1982.

Shabert, Judy, and Nancy Ehrlich. *The Ultimate Nutrient: Glutamine* (Garden City Park, N.Y.: Avery Publishing Group, 1994).

Shahami, Khem M., and Armadu D. Ayebo. "Role of Dietary Lactobacilli in Gastrointestinal Microecology," *The American Journal of Clinical Nutrition,* vol. 33, pp. 2448–57, November 1980.

Tateishi, N., et al. "Relative Contributions of Sulfur Atoms of Dietary Cysteine and Methionine to Rat Liver Glutathione and Proteins," *Journal of Biochemistry,* vol. 90, pp. 1603–10, 1981.

Tyler, Varro E. *Herbs of Choice* (Binghamton, N.Y.: Pharmaceutical Products Press, 1994).

CHAPTER 3

GLUTATHIONE AND ENVIRONMENTAL ILLNESS

In my practice as a chiropractor and nutritionist, I often encounter patients who have come to me after a lengthy period—sometimes years—of steadily worsening health. These patients have seen many different medical doctors in a vain search for help. Often they are taking half a dozen different prescription drugs. But because their symptoms fluctuate and often include insomnia, depression, an inability to concentrate, digestive upsets, and other vague complaints, and because their symptoms can't be pinned down by standard medical tests and do not respond well to standard medical management, these patients have been told that their problems are all in their heads.

These patients aren't imagining things. They are physically sick from their exposure to toxic substances in their environment. Fortunately, once they are finally diagnosed correctly, effective treatment can begin.

One of my patients, Mary, was sure her mysterious illness had a physical cause, though her doctor at first blamed it all on "stress." Mary was a high school swimming coach who enjoyed good health, swam regularly for exercise, and had a happy family life. After many years at the high school level, she took a job as swim coach at a local college.

Within eight weeks of starting her job at the college, she began to feel fatigued and depressed. Her work was no longer a joy. She found herself sleeping much more, which meant she spent less time with her family.

Then her headaches began. Mild and occasional at first, they soon grew severe and constant; sometimes they were so bad that Mary felt confused and disoriented.

Mary's doctor was also confused. Mary was obviously getting sicker, but he couldn't find anything medically wrong. He thought that the stress of a new, high-pressure job might be causing Mary to feel depressed. Mary, however, insisted that she loved her new job and enjoyed its challenges. When all else had been ruled out, Mary's doctor began to suspect environmental illness based on her exposure to the chlorine used in the swimming pool. But why would Mary suddenly be sick from chlorine? For years at her previous job she'd spent hours every day next to a chlorinated swimming pool and never had any trouble. Just to be sure, the college brought in environmental engineers to study the swimming facility. They found a faulty exhaust system that was allowing chlorine gas to build up in the air instead of being vented outside the building. Mary was inhaling so much chlorine that her body couldn't filter it all out. Instead, it built up in her system and made her sick.

The exhaust system was fixed quickly, but Mary needed a leave of absence to regain her health. She came to me for nutritional counseling to help speed her recovery. We put her on a program rich in the nutrients that help the body detoxify, with the goal of up-regulating her liver function and clearing the toxins from her system faster. Mary took supplemental GSH, selenium, and lipoic acid, along with antioxidant vitamins. She also drank eight glasses a day of pure water and ate a low-fat diet with plenty of fresh fruits and vegetables. In six weeks, Mary was happily back at work.

THE TOXIC ENVIRONMENT

When people are regularly exposed to toxic chemicals, sooner or later they will become ill. A good example of how even small amounts of allegedly safe and even useful chemicals can combine to cause illness is what doctors have come to call Gulf War syndrome. Military personnel who served in the Middle East in 1991 as part of this conflict were exposed to a number of drugs, insecticides, and toxins such as the fumes from burning oil wells. Many later reported a wide variety of symptoms, including unexplained rashes, fatigue, headaches, inability to concentrate, and so on. To their credit, military officials investigated carefully but concluded that about a third of the veterans with symptoms—some ten thousand soldiers—were not suffering from any identifiable illness. Their symptoms were said to be psychological or of no known origin.

In 1996, however, work by university researchers showed that a combination of two chemicals widely used during the conflict could indeed cause the symptoms of Gulf War syndrome. Insecticides containing organophosphates, which inhibit production of the enzyme cholinesterase, were used extensively. To protect the soldiers against the effects of possible chemical warfare, the troops were also given a drug called pyridostigmine. The two substances in combination, even at fairly low levels, can cause nerve and brain damage in some individuals.

Why did only some Gulf War veterans develop symptoms? The pyridostigmine blocks enzymes that would ordinarily cope with the organophosphates, allowing the insecticide to reach the nerves and brain. People who naturally have low enzyme levels will be more effectively blocked by the pyridostigmine and therefore will be less able to fight off the effects of the organophosphates.

Most of us, of course, lead safe, comfortable lives and don't need to be protected against the effects of chemical warfare. Or do we? Look around the room you're in as you read this. Environmental toxins surround you. You might be wearing formaldehyde-containing polyester clothing that was washed with chloroform-containing fabric softener; polyester might be in the furniture, curtains, or cushions in the room. Foam rubber in a cushion or mattress could contain formaldehyde; it could also be in paneling on the wall, in the insulation behind the wall, in the carpet, or in the padding under the carpet. The furnace heating the room could be emitting oil or gas vapors or carbon monoxide, while air pollution and exhaust fumes laden with heavy metals, dioxin, benzene, sulfur dioxide, and other poisons could be coming in through the open window. You're breathing in toxins and absorbing them through your skin—and we have yet to discuss the toxins that are in the food you eat and the water you drink.

Environmental toxins surround all of us all the time. Indeed, without making some fairly radical changes in your lifestyle, you can hardly avoid them. In the long run, environmental toxins are bad for everyone. As we'll discuss in later chapters, long-term exposure to environmental toxins can lead to cancer, heart disease, autoimmune diseases, and other serious health problems. In the short term, environmental toxins can also have serious effects. A brief exposure to an individual toxin such as benzene (often found in dry-cleaning shops) can cause a measurable rise in your blood levels of the toxin, even if you don't have any immediate symptoms.

Let's say you stop off at the dry-cleaning shop and are briefly exposed to benzene. You then put gas in your car at the service station, again exposing yourself to benzene (and other toxins) in the gas fumes. Then you spend an hour visiting a friend who has just painted her living room—benzene again. When you finally get home, you're tired, irritable, and have a headache. You might vaguely attribute it to stress; you might even make a connection between the smell of paint and your headache. But the headache is gone in a few hours and you don't think about it anymore.

What happens if your benzene exposure continues? At first, probably nothing beyond the occasional mild symptoms already described. Over years of exposure, however, your chances of developing life-threatening leukemia are increased. If you are continuously exposed to benzene—if you work in or live near the dry-cleaning shop, for example—your risk is markedly higher.

If your benzene exposure is combined with exposure to some other toxin(s), what happens then? The truly frightening answer is that nobody really knows. There are tens of thousands of chemicals in common industrial use today, and most have never been tested for their effects on humans. Very few indeed have been tested for their effects in combination. Fabric softeners, for example, contain nine different chemicals that the U.S. Environmental Protection Agency (EPA) says can cause central nervous system damage or cancer. Each one of these chemicals is dangerous in itself; no one knows how much more dangerous they are together.

ENVIRONMENTAL ILLNESS

Most healthy people can handle a steady, moderate dose of xenobiotics (foreign chemicals), at least for a while. Your body's metabolic pathways manage to keep abreast of the toxic load and eliminate it from your system. Even so, continuous exposure to environmental toxins can eventually lead to the accumulation of a toxic load that can cause chemical hypersensitivity and environmental illness (EI). If you are badly nourished or unhealthy for some reason, xenobiotics could affect you more severely. Some people are naturally more susceptible to certain xenobiotics because of genetic variations in their liver enzymes and other aspects of their body chemistry. In a less toxic world, these people would be just as healthy as anyone else. Unfortunately, in our modern world they are more likely to get sick.

In the last chapter, you learned how your liver detoxifies your body and removes dangerous wastes. You also learned how important it is to keep your detoxification pathways working efficiently. It should be very clear, then, that your detox pathways will not work efficiently if they are overwhelmed with removing large amounts of environmental toxins from your system. When the toxins aren't removed efficiently, they can build up in your tissues, especially in your fatty tissues and the fatty sheaths surrounding your nerve cells; they can also damage your endocrine and immune systems. Cellular damage that can later lead to cancer can be one serious result. More immediately, a large backlog of xenobiotics often causes nervous system and brain symptoms such as confusion, depression, inability to concentrate, headaches, tremors, weakness, and fatigue. When dealing with environmental illness, however, it's important to remember that symptoms can appear anywhere, not just in the nervous system. In addition, symptoms can vary widely from person to person and, in the same person, from day to day or even from hour to hour.

Figure 3.1 lists the most common symptoms of environmental illness. You may notice that some of the symptoms seem contradictory—insomnia and excessive sleepiness, for example. Some medical practitioners claim that the symptoms of EI are so broad that they are meaningless, and that therefore the illness somehow

Figure 3.1

COMMON SYMPTOMS OF ENVIRONMENTAL ILLNESS

Nervous System/Brain:

depression

fatigue

poor memory

inability to concentrate

headaches

dizziness

"spaciness," "fogginess," or
  feelings of unreality

insomnia

excessive sleepiness

irritability

restlessness

lethargy

panic attacks

behavior or
  personality changes

lack of coordination

claustrophobia

Muscles and Bones:

fatigue

joint pain

swollen joints

arthritis

aching muscles

muscle twitches and
  spasms

muscle weakness

lack of coordination

Digestive System:

loss of appetite

food cravings, especially for
sugar

nausea

vomiting

abdominal pain or
  cramps

constipation

diarrhea

weight loss

Heart and Lungs:

heart arrhythmias

chest pain or tightness

wheezing

shortness of breath

Nose, Mouth, Sinuses, Throat:

inability to taste foods well

inability to smell well

sore throat

persistent hoarseness

laryngitis

nasal or sinus burning

| COMMON SYMPTOMS OF ENVIRONMENTAL ILLNESS (cont.) |

Skin and Nails:

acne	sensitivity to sun
skin rashes	increased sweating
dry skin	discolored or deformed nails
flushing	slow healing of cuts

Menstrual:

severe PMS	severe menstrual symptoms

Immune System:

swollen glands in neck and armpits	mononucleosis or Epstein-Barr infection
chronic fatigue syndrome	yeast infections
	worsening food or inhalant allergies

Source: Sherry A. Rogers, M.D., *Tired or Toxic?* (Syracuse, N.Y.: Prestige Publishers, 1990; 800-846-6687).

doesn't exist. In fact, the contradictory symptoms indicate how variable and changeable environmental illness symptoms can be. The reason for the variability is simply that the total toxic load of people with EI varies from day to day—so their symptoms do too.

Another result of toxic buildup can be chemical hypersensitivity: increased sensitivity not just to the toxin in question but also to other toxins that weren't a problem previously. This can lead to a downward spiral of increasingly severe reactions (multiple chemical sensitivity) to more and more substances. Dr. Sherry A. Rogers, a leading environmental physician, calls this the spreading phenomenon.

ARE YOU ENVIRONMENTALLY ILL?

Many of the symptoms of environmental illness are similar to those of other illnesses. It is very important to rule out other possible causes of the symptoms before diagnosing EI. For example, chest tightness combined with weight loss, wheeziness, and fatigue could indicate lung cancer.

In my experience, many patients with suspected environmental illness also have food allergies and often inhalant allergies as well. It's essential to discover and treat the allergies as well as the environmental illness. Discuss your allergies with your health care practitioner and do everything you can to get them under control. Once your body is no longer trying to cope with allergic responses, it can devote more energy to dealing with environmental toxins.

Whenever environmental illness is suspected, leaky gut syndrome, dysbiosis, and sluggish liver must also be suspected and treated if necessary. Leaky gut syndrome, which allows large, undigested food molecules to be absorbed into your body through the small intestine, can trigger responses very similar to environmental illness. As discussed in chapter 2, detecting and treating leaky gut syndrome can help reduce the toxic load on your liver from the foods you eat. This will allow your liver to detoxify environmental poisons more efficiently. Your own internal environment could be producing toxins if you have dysbiosis. Again, your liver could be so busy coping with internal toxins that it has few enzymes and little glutathione left over to tackle xenobiotics—or vice versa. Poor liver function is another possible cause of environmental illness. If your Phase I and Phase II activity are out of balance and are not detoxifying your body effectively, toxins will build up (see chapter 2).

Low zinc, selenium, and magnesium levels can sometimes make you more susceptible to environmental illness. Not enough, and your enzyme production will be too low to be effective and you could build up a backlog of toxins in your system. Your body needs zinc to make many of the enzymes that help remove toxins. Selenium is a vital component of the enzyme glutathione peroxidase, needed to recycle the glutathione in your body. In addition to capturing free radicals, glutathione is an extremely effective remover of toxic substances. Inadequate levels of selenium can lead to inadequate levels of glutathione and a dangerous build up of toxins in your tissues. Magnesium plays an important role in removing toxins such as ammonia (a byproduct of cell metabolism) from your system; it's also needed for your body to make many proteins. Again, a shortage of magnesium allows a toxic backlog to accumulate. Blood tests can reveal low levels of zinc or magnesium; testing for glutathione level (see chapter 2) can reveal low levels of selenium and glutathione.

Poor diet or heavy metal poisoning are possible causes for low levels of these essential minerals. A diet that is high in fat, low in fiber, and high in processed foods can lead to nutritional deficiencies, including low levels of important minerals. (Heavy metal poisoning is discussed later in this chapter.) Often, changing your diet to include more fresh fruits and vegetables and plenty of legumes (beans and peas) and whole grains can help improve your health. To make sure you're getting adequate amounts of these minerals, I suggest you also take a high-quality daily multivitamin from a reputable manufacturer such as Solgar, Nature's Way, or Ultimate Bionetics.

There are a few laboratory tests your health care provider can do that can help determine if you are environmentally ill. Sometimes blood tests reveal that your total white blood cell count and your lymphocyte count are low and your antibody levels are high. This suggests that your body is producing autoimmune antibodies in response to the presence of xenobiotics. As we'll discuss in more detail in chapter 6, high levels of interleukins in your blood could also suggest an autoimmune response to toxins. A blood test that looks at the levels of formic acid in your blood can show if formaldehyde is the culprit. A functional liver profile, oxidative stress level, and testing for intestinal permeability (see chapter 2) could reveal poor liver function. Other lab tests may also help diagnose EI, including blood tests for your levels of globulin, albumin/globulin ratio, cholesterol, total protein, and liver and kidney enzymes. A few laboratories can do other specialized testing to determine the presence of some dangerous xenobiotics, but discuss these with your doctor before you go ahead. (See Appendix A for the names of reliable labs.) These tests are expensive and may not really be needed—those described here, along with a careful assessment of your symptoms, should be enough for a diagnosis of environmental illness.

CURING ENVIRONMENTAL ILLNESS

Once you've discovered that you have EI, you are already well on the way to curing it. You now know if you have any allergies, if you suffer from leaky gut syndrome or dysbiosis, and if your levels of GSH, zinc, selenium, and magnesium are low. You also now know the steps to take to deal with these aspects of your health.

Reducing Your Exposure

The important next step is to reduce your exposure to xenobiotics. Although this is not always easy, you can take a number of surprisingly simple steps that can sharply lower your exposure. You'll feel better, and your family will too. Unfortunately, I have space here for only a short discussion. For excellent guides to living a more chemical-free life, I recommend *Nontoxic, Natural, and Earthwise*, by Debra Dadd, and *The Safe Shopper's Bible*, by David Steinman and Samuel S. Epstein, M.D.

First, check your home for chemical fumes. Just looking under the sink could reveal a major source of your environmental illness. Get rid of all the insecticides, mildew removers, mothballs, scented laundry detergents, fabric softeners, perfumed cleansers, and other chemical-laden products you see there. There are plenty of readily available, nontoxic alternative products that work as well if not better. Check other places in your house as well. Those air fresheners may make the room smell nice, but they could be poisoning you. All those fertilizers, herbicides, and pesticides in the garage are poisoning you and the planet. Learn to garden organically.

Other sources of chemical fumes in the home could include foam rubber, foam insulation, scented soaps, perfume, cosmetics, cigarette smoke, automobile exhaust from an attached garage, and gas or oil fumes from your heating system. The many and various plastics in your home could be outgassing, or emitting toxic fumes that are making you sick. Be especially wary of anything that has a "plastic" odor.

It's also possible that your tap water contains chemicals such as chlorine that have been added to your municipal water supply. The water could also contain lead or copper it picks up from the pipes; in areas with very soft water, cadmium can leach from your pipes. Rusty pipes can give you unwanted iron buildup in your system. To find out what is being added to your water before it reaches your house, ask your local water supplier for a water-monitoring report. You can also have your tap water tested for contaminants and bacteria by any local testing laboratory or by the mail-order water testing companies listed in Appendix A.

I strongly urge you to install water filters or purchase pure filtered water in glass containers. I find the inexpensive Water Pure filter system, which easily attaches to any water tap, to be very effective. (For more information, call 800-313-7873.)

The foods you eat can contain xenobiotics such as preservatives, insecticides, and herbicides. Read ingredient labels carefully for the presence of preservatives and additives such as BHT or sodium nitrite. Wash fresh fruits and vegetables before eating. The best way to do this is to mix two tablespoons of distilled vinegar in a gallon of filtered tap water. Soak the fruits and vegetables for fifteen minutes, then drain, rinse, and dry. Even better, purchase only organically grown foods.

Chemical exposure in your workplace is another matter, since you have less control over it. If you work in a sealed building with no natural ventilation, you may be getting large doses of outgassed xenobiotics such as formaldehyde and benzene from carpets, furnishings, and plastics every day—a situation known to occupational health experts as sick building syndrome. Cigarette smoke from your coworkers, industrial cleaning solutions, and fumes from photocopiers and office supplies could be adding to your xenobiotic load. In fact, chemicals in the workplace are one reason that people with environmental illness often get sicker as the week goes on, recover somewhat over the weekend or while on vacation, and then get sick again as soon as they go back to work— and why they are sometimes accused of just being lazy.

Of course, if you work in an industrial environment such as a factory, you are probably being exposed to all sorts of potentially dangerous chemicals. Agricultural and construction work can also expose you to xenobiotics. Follow all safety procedures and wear protective clothing and other gear as needed.

If your workplace exposure is high, you can take some simple steps to reduce it. Ask your coworkers to observe smoking bans, for example. Do what you can to avoid exposure to chemicals from office supplies—perhaps you could swap those parts of your job with someone else or have your desk moved. You could also place a spider plant or two on your desk; these plants naturally absorb air-borne toxins such as benzene. If building ventilation is poor, try discussing it with the maintenance supervisor.

Dietary Measures

The next step in curing environmental illness is eliminating the accumulated toxins from your system. Many of my patients benefit greatly from a complete program of metabolic cleansing, using Ultra Clear or Bio Detox Powder as outlined in chapter 2. The program helps them eliminate their backlog of toxins and get their liver working more effectively. Some of my patients find that they feel worse for the first few days of the program as their body eliminates poisons; once they get through the initial discomfort, however, they usually start to feel markedly better. Remember, metabolic cleansing programs should be done only under the supervision of a health care provider.

Metabolic cleansing alone will not eliminate all the toxins, however. Once you've completed the program, I recommend a high-fiber, low-fat diet with plenty of fresh fruits and vegetables and lots of whole grains. If your cholesterol levels are normal and if you do not have coronary artery disease or heart disease, I also recommend two to four eggs a day for six months. Eating eggs may help repair the damage xenobiotics have done to your body, particularly to your nerve and brain cells. As we'll discuss more in chapter 9, eggs may stimulate your body to produce apoproteins, an essential component of the lipoproteins your body uses to transport cholesterol. Since you need cholesterol to repair nerve damage, increasing your apoprotein level may speed your recovery. Apoproteins are also helpful for repairing liver damage.

Keeping your body well-hydrated speeds the elimination of toxins. Drink six to eight 8-ounce glasses of pure water every day.

Another important part of treatment for xenobiotics is daily aerobic exercise. Exercise stimulates your body and helps improve depression and other mental symptoms of environmental illness. Also, sweating is another way your body eliminates toxins.

My EI patients benefit greatly from adding antioxidant supplements, vitamins, and minerals, but without question the supplement that helps them most is glutathione. Some patients are helped by taking glutathione directly, while others benefit more from supplements of N-acety-L-cysteine. You may have to experiment a bit to find which is best for you. No matter which form you prefer, glutathione can be fantastically helpful for eliminating toxins. Glutathione circulates throughout your body, corralling unwanted toxins wherever it finds them—not just in your liver.

Daily Supplements for EI

Here's what I suggest as your daily supplements for battling environmental illness:

NAC	600 mg
GSH	250 mg
Lipoic acid	200 mg
Selenium	250 mcg
Magnesium	400 mg
Zinc	30 mg

If you prefer, take the above supplements in a convenient multi form. I recommend Glutaplex or Glutathione 250 Master Formula from Douglas Labs; take three capsules daily. Hepatox from Biometrics is another good formula; follow the dosage instructions on the bottle.

Once you have begun your treatment program, you will almost certainly feel markedly better in just a few weeks. Don't stop the program. Many xenobiotics find their way into your body's fatty tissues, so it could be months or possibly years before they are all eliminated; in the meantime, they could continue to cause symptoms. Ask your health care provider to help you evaluate your body composition and determine your ratio of lean body mass to fat. If you are very lean, you will probably eliminate most of your toxic load quickly. If you have a high ratio of body fat to lean mass (this does not necessarily mean you are obese), you will store more toxins and will probably take longer to eliminate your full toxic load.

Metabolic cleansing programs are not weight-loss diets. If you have EI symptoms and are also overweight, the most important objective is to reduce your xenobiotic load. Once you've done that, go on a sensible diet that will help you lose weight slowly. When you diet, toxins that have been stored in your fatty tissue are released. If you lose weight slowly, your liver will be able to cope with the load. If you lose weight rapidly from a crash diet, you will release a lot of toxins that could overburden your liver.

Because you may be naturally more susceptible to xenobiotics, you will always have to protect yourself against them. In addition, your bout with environmental illness may have made you permanently hypersensitive to xenobiotics. You should continue to be very vigilant about your exposure to toxins. I also strongly suggest continuing a maintenance program of glutathione supplementation. A good choice would be two capsules daily of GSH 250 Master Glutathione Formula. I also suggest taking a weekly drink at breakfast of Bio Detox Powder or Ultra Clear.

HEAVY METAL POISONING

Just as we are constantly surrounded by a variety of chemical toxins, we are also constantly exposed to microscopic bits of metals such as aluminum, arsenic, cadmium, lead, mercury, and nickel. (See Figure 3.2 for some possible sources of exposure.) Often called heavy metals (although, technically speaking, only lead and mercury are truly heavy, in the sense of having high atomic weights), these substances can cause serious illness and long-term damage to your health. In many cases, heavy metals interfere with your body's normal production of the many enzymes it uses to regulate your metabolism. For example, arsenic, which differs chemically from selenium by just one atom, can displace selenium in your body by combining with sulfur-containing enzymes such as glutathione. When your body can't produce glutathione correctly, the result is massive free radical damage. In other cases, heavy metals interfere with your ability to absorb and use other important minerals such as calcium and magnesium. Heavy metals

Figure 3.2

COMMON SOURCES OF HEAVY METAL POISONING

Metal	Source
Aluminum	foil; cans; deodorants; buffered aspirin; some antacids; aluminum utensils and cookware; baking powder
Arsenic	insecticides; herbicides; paints; some ceramics
Cadmium	paints; metal plating; colored plastics; gasoline; fertilizers; fungicides; cigarette smoke; batteries; solder
Lead	chips or dust from lead-based paint; leaded gasoline; solder; inks and dyes; pottery glazes; batteries
Mercury	silver amalgam dental fillings; paints; fungicides; fabric softeners; floor waxes and polishes; photo film; some plastics; latex house paints before 1990; batteries
Nickel	metal alloys in costume jewelry; cigarette smoke

Note: All these metals have many industrial uses and are often found in the soil, air, and water of industrialized areas.

often affect the brain, leading to symptoms such as depression, memory loss, anxiety, and learning disabilities. These symptoms are often mistaken for emotional or mental problems. Once heavy metal poisoning has been diagnosed, treatment with glutathione and cysteine can be very helpful, miraculously curing many so-called mental illnesses and physical ailments.

Aluminum

Aluminum is one of the most common and useful elements on earth, with many industrial, medical, and household uses. Too much aluminum, however, can lead to bone and brain disorders. Excessive aluminum interferes with your body's ability to absorb and utilize calcium, fluoride, magnesium, phosphorus, and selenium. If aluminum keeps you from utilizing calcium and phosphorus properly, the result could be osteoporosis. Excessive aluminum could also cause memory loss, dementia, and learning difficulties. As we'll discuss in chapter 5, excessive aluminum may also be associated with Alzheimer's disease.

Arsenic

Although arsenic is well-known as a deadly poison, your body actually needs it in very tiny amounts. When you accumulate more than the trace needed, however, serious symptoms develop. Numbness, tingling, and burning sensations in the hands and feet are early signs of arsenic poisoning. Other symptoms include muscle weakness, headaches, fatigue, and dermatitis. Serious liver and kidney damage and even death can result from the cumulative exposure to quite small amounts of arsenic. Most of us can avoid this dangerous poison by avoiding pesticides and herbicides. However, arsenic is also a byproduct of copper smelting, so if you work in or live near or downwind from a copper smelter or a manufacturing plant that uses copper, you could be at real risk.

Cadmium

Cadmium is naturally found in many foods, including shellfish, fish, coffee, poultry, grains, and dairy products. The amounts are so small, however, that your body has never evolved any pathways for excreting cadmium; if your exposure is high, cadmium can accumulate to toxic levels. In modern society, cadmium is widely used in paints, plastics, and other industrial applications. It is also a byproduct of zinc refining. For most people, however, the most serious cadmium exposure comes from cigarette smoke. If you are pregnant, cadmium from cigarettes can seriously damage your unborn baby. If you smoke, stop.

Cadmium can cause anemia, muscle pain, and high blood pressure. It may also increase your susceptibility to certain cancers, particularly lung cancer and prostate cancer. Learning disabilities and low intelligence can also be results of cadmium toxicity.

Lead

The ill effects of lead poisoning, especially on small children, have been well known for decades. Lead atoms have a great affinity for sulfur atoms, including the crucial sulfur atom in the amino acid cysteine. Exposure as low as one part per million will cause the sulfur in cysteine to combine with the lead, instead of combining properly with other amino acids to form enzymes—including glutathione—in the body. The enzymes malfunction, which in turn causes the wide range of symptoms that characterize lead poisoning. Because lead must be made by refining galena ore, a process that was developed only about five thousand years ago, humans have never evolved a way to remove it efficiently from their bodies. The only way to treat lead poisoning is through chelation therapy, which uses a drug that binds strongly to lead atoms and helps excrete them from the body. The therapy is only partially effective.

Despite this frightening knowledge, dangerously high lead levels continue to be a major public health problem. According to the Centers for Disease Control, nearly 9 percent of American children between the ages of one and five have potentially hazardous blood levels higher than 10 mcg per deciliter. These children, many of them from poor families, are at great risk for learning disorders, behavior problems, and permanent brain damage.

Lead poisoning can cause a wide range of emotional and behavioral problems and has often been linked to criminal behavior. In 1996 the results of a study of boys attending public school in Pittsburgh found a strong correlation between the amount of lead in their bones and their levels of aggression and delinquency. In short, the boys with the highest lead levels were the most likely to be delinquent and aggressive.

Toxic lead levels can also show up in children and adults as poor scores on intelligence tests, poor learning ability, hyperactivity, depression, and poor memory. Physical symptoms of lead poisoning may include anemia, free radical damage, seizures, and increased susceptibility to infection. Another recent study has shown that even very low levels of lead can cause high blood pressure and kidney damage in adults. Currently, standards set by federal agencies such as the Occupational Safety and Health Administration (OSHA) say that adult blood levels of 40 to 50 micrograms per deciliter are acceptable; this figure may need to be revised sharply downward. I strongly urge you not to wait for the government to act—do everything you can to reduce your family's exposure to lead. See Figure 3.2 for possible sources of lead exposure in your home. Old paint is the most likely source, but removing it by sanding will expose you to tiny particles of lead. Overpaint with a high-quality interior paint instead.

Mercury

The old expression "mad as a hatter" comes from the old practice of using mercury to make the felt used in hats. In earlier times, many hatters developed mercury poisoning, which can indeed lead to disturbing psychological symptoms such as dementia, anxiety, poor memory, and irritability. Mercury can also affect the nerves, causing tremors, seizures, severe headaches, and loss of coordination. Unpleasant mouth symptoms such as excessive salivation and ulcerations are also symptoms of mercury poisoning. Today far less mercury is used in making felt and hats, and safety measures in manufacturing operations are much more effective. Very few people will ever experience serious mercury poisoning. Low levels of mercury poisoning, however, often still occur.

Mercury is found in many household products such as floor waxes and fabric softeners; it's also used in some plastics and adhesives. However, most people are exposed to mercury through the silver amalgam fillings in their teeth. The mercury in the mixture can slowly leach out of the filling and into your body. If you just have one or two fillings, this may not present a problem, but if you have a lot, you could gradually build up mercury to toxic levels. The American Dental Association continues to claim that there is no reason to replace your old fillings, but common sense suggests otherwise. Several of my patients got rid of nagging health problems such as psoriasis and frequent headaches by having their old silver amalgam fillings replaced with safer (and more natural-looking) ceramic fillings.

Until 1990, when it was banned by the FDA, mercury was also an ingredient in latex house paints. If your house was last painted, inside or out, before 1992 (manufacturers were allowed to sell their stock on hand after the ban went into effect), you may be getting some mercury exposure from the walls around you. If you suspect mercury-containing latex paints were used in your home, it's time to repaint.

Many over-the-counter drugs contain small amounts of mercury compounds; they are also found in some contact lens solutions and in some vitamin tablets. Read the labels carefully, and purchase vitamins only from respected manufacturers such as Solgar.

Nickel

Nickel is another metal that you need in trace amounts. It is not very well absorbed by your body through the intestines; in fact, your body rapidly excretes virtually all the nickel it takes in, so you can't really get nickel poisoning. However, nickel that enters the lungs through cigarette smoke is not excreted. Instead, it stays in the lungs, where it may later cause cancer.

Most nonsmokers encounter nickel as an ingredient in the metal alloys of costume jewelry. Many people find that nickel irritates their skin, causing a rash (contact dermatitis) wherever the nickel touches them. The simplest solution is to stop wearing the jewelry.

DO YOU HAVE HEAVY METAL POISONING?

Heavy metals tend to accumulate in your bones and in the spaces between your cells, so they don't show up in blood or urine tests unless your levels are dangerously high. To detect heavy metals, your doctor may do a test called provocative chelation. The patient is given a safe drug that attaches to the metal (chelates) and pulls it into the blood. Once the metal is in your bloodstream, it will soon be excreted in your urine, where it can be detected and measured.

Another, somewhat controversial way to detect heavy metal poisoning is through hair analysis. The problem is that your hair normally contains some small amounts of metals, which can make the test results misleading. More reliable lab techniques for hair analysis have recently been developed, however. The results from today's best labs are more accurate and may reveal hidden cases of heavy metal poisoning.

TREATING HEAVY METAL TOXICITY

As with the environmental toxins discussed above, the first step in treating heavy metal toxicity is to reduce your exposure. If you have severe heavy metal poisoning, your doctor may prescribe safe and effective chelating drugs such as d-penicillamine or DMSA. If your toxicity is low, I suggest you try using glutathione or cysteine, along with other supplements and dietary changes, to remove the metals from your system.

Glutathione is your body's primary defense against toxic metals. As glutathione circulates in your body, it binds to metals, particularly lead, mercury, and arsenic, and lets them be safely excreted from the body. Because glutathione circulates not just in the blood but in the spaces between the cells where heavy metals are stored, it can help to chelate them into the blood and then out of the body. Of course, the antioxidant effect of glutathione helps prevent cellular damage from the heavy metals.

Vitamin C in large doses (up to 5,000 mg a day) can also help chelate heavy metals and prevent free radical damage. But discuss high doses with your health care provider before you try this; large doses of vitamin C can cause diarrhea. The minerals calcium, magnesium, and zinc are also heavy metal chelators.

Garlic may also help. Like cysteine, the basic building block of glutathione, garlic contains sulfur, and like cysteine, it has a chelating effect on heavy metals. I recommend odorless garlic capsules or tablets such as those made by Kyolic, since they are convenient and don't give you bad breath. One garlic capsule is about the equivalent of one large raw garlic clove.

Daily Supplements for Heavy Metal Toxicity
If you are suffering from low-level heavy metal toxicity, here's what I recommend:

GSH	1,500 mg
NAC	1,500 mg
Lipoic acid	400 mg
Selenium	250 mcg
Calcium	1,000 mg

Magnesium	500 mg
Zinc	50 mg
Vitamin C	5,000 mg
Garlic	3 capsules daily

For best results, spread out the doses over a day. In addition, take a good multivitamin from a reputable manufacturer such as Solgar or Nature's Way.

In addition to taking chelating supplements, I suggest a high-fiber diet that contains lots of beans and peas and also high-pectin fruits such as apples and pears. These foods can help bind heavy metals in your diet and keep them from being absorbed into your body through the intestines. The chlorophyll in green leafy vegetables such as kale and Swiss chard can also help you excrete metals. In addition to eating plenty of fresh vegetables, try taking a chlorophyll-containing greens formula such as Phytogreens. I also suggest drinking six to eight 8-ounce glasses of pure water daily. As discussed above, if your cholesterol levels are normal, eating two to four eggs a day may help repair the nerve and brain damage done by heavy metals.

Removing heavy metals from your body takes a long time, often six months or longer. It may be several weeks before you start to feel any effects.

My patients almost always start to feel much better as soon as they start ridding their bodies of environmental toxins and restoring their normal metabolic functions. They feel great when the process is complete. But for some patients, even though they feel better now, help may have come too late. Prolonged exposure to environmental toxins is one of the leading causes of cancer.

REFERENCES

Broughton, A., J. D. Thrasher, and R. Madison. "Chronic Health Effects and Immunological Alterations Associated with Exposure to Pesticides," *Comments Toxicology,* vol. 4, pp. 59–71, 1990.

Chaudhari, A., and S. Dutta. "Alterations in Tissue Glutathione and Angiotensin Converting Enzyme Due to Inhalation of Diesel Engine Exhaust," *Journal of Toxicology and Environmental Health,* vol. 9, no. 2, pp. 327–37, 1982.

Dadd, Debra. *Nontoxic, Natural, and Earthwise* (Los Angeles: J. T. Tarcher, 1990).

Dager, S. R., et al. "Panic Disorder Precipitated by Exposure to Organic Solvents in the Workplace," *American Journal of Psychiatry,* vol. 144, no. 8, pp. 1056–58, August 1987.

Feldman, R. G., N. C. Ricks, and E. L. Baker. "Neuropsychological Effects of Industrial Toxins," *American Journal of Industrial Medicine,* vol. 1, pp. 211–27, 1980.

Gershon, S. "Psychiatric Sequelae of Chronic Exposure to Organophosphorus Insecticides," *Lancet,* pp. 1371–74, June 24, 1961.

Hsu, J. M. "Lead Toxicity as Related to Glutathione Metabolism," *Journal of Nutrition,* vol. 3, pp. 26–33, 1981.

Hu, H., A. Aro, M. Payton, et al. "The Relationship of Bone and Blood Lead to Hypertension: The Normative Aging Study," *Journal of the American Medical Association,* vol. 275, no. 15, April 17, 1996.

Husain, S., and D. Dunlevy. "Possible Role for Glutahione in Phencyclidine Toxicity and Its Protection by N-acetylcysteine," *The Pharmacologist,* vol. 243, no. 3, 1982.

Jensen, L. S., and D. V. Maurice. "Influence of Sulfur Amino Acids on Copper Toxicity in Chicks," *Journal of Nutrition,* vol. 109, pp. 91–97, 1979.

Kawata, M., and K. T. Suzuki. "The Effect of Cadmium, Zinc, or Copper Loading on the Metabolism of Amino Acids in Mouse Liver," *Toxicology Letters,* vol. 20, pp. 149–54, 1984.

Kendall, Julia. "Fabric Softeners = Health Risks from Dryer Exhaust and Treated Fabrics," *Townsend Letter for Doctors and Patients,* December 1995.

Kim, R., A. Rotnitzky, D. Sparrow, et al. "A Longitudinal Study of Low-Level Lead Exposure and Impairment of Renal Function: The Normative Aging Study," *Journal of the American Medical Association,* vol. 275, no. 15, April 17, 1996.

Martinez-Torres, C., E. Ramono, and M. Layrisse. "Effect of Cysteine on Iron Absorption in Man," *American Journal of Clinical Nutrition,* vol. 34, pp. 322–27, 1981.

Needleman, H. L., J. A. Riess, M. J. Tobin, et al. "Bone Lead Levels and Delinquent Behavior," *Journal of the American Medical Association,* vol. 275, no. 5, February 7, 1996.

Nutrition Reviews, "Effects of Lead on Glutathione Metabolism," vol. 39, no. 10, pp. 378–79, 1981.

Randolph, Theron G., and Ralph W. Moss. *An Alternative Approach to Allergies* (New York: Lippincott/Crowell, 1980).

Randolph, Theron G. "Sensitivity to Petroleum, Including Its Derivatives and Antecedents," *Journal of Laboratory Medicine*, vol. 40, pp. 931–32, 1952.

Rea, W. J., et al. "Food and Chemical Susceptibility After Environmental Chemical Overexposure," *Annals of Allergy*, vol. 41, no. 2, pp. 101–7, August 1987.

Rogers, Sherry A., M.D. *Chemical Sensitivity* (New Canaan, Conn.: Keats Publishing, 1995).

Rogers, Sherry A., M.D. "Diagnosing the Tight Building Syndrome," *Environmental Health Perspectives*, vol. 76, pp. 195–98, 1987.

Rogers, Sherry A., M.D. *The E.I. Syndrome* (Syracuse, N.Y.: Prestige Publishers, 1990).

Rogers, Sherry A., M.D. *Tired or Toxic?* (Syracuse, N.Y.: Prestige Publishing, 1990).

Steinman, David, and Samuel S. Epstein, M.D. *The Safe Shopper's Bible: A Consumer's Guide to Nontoxic Household Products, Cosmetics, and Food* (New York: Macmillan, 1995).

Thrasher, J. D., A. Broughton, and R. Madison. "Immune Activation and Autoantibodies in Humans with Long-Term Inhalation Exposure to Formaldehyde," *Archives of Environmental Health*, vol. 45, pp. 217–23, 1990.

GLUTATHIONE AND CANCER PREVENTION

Of the million or so cases of cancer diagnosed every year in the United States, many—perhaps most—could have been prevented. How? With two things: more antioxidants and no cigarettes.

You already know the many health dangers of cigarette smoking. If you haven't quit already, do it now. What you may not know is that, by the most conservative estimate, some 35 percent of all cancers are directly related to your diet. Some respected researchers maintain that environmental factors, including diet, are at the root of some 80 percent or more of all cancer cases.

Cancer is a complex disease that can have many different causes. There are more than one hundred different types of cancer, and we still have a lot to learn. Today, however, many cancer researchers believe that a diet rich in antioxidant nutrients could significantly reduce your risk of cancer by reducing your risk of cell damage from free radicals.

THE CANCER CASCADE

The vitamins and micronutrients most closely associated with reduced cancer risk are also the most powerful antioxidants. The vitamins include mixed carotenes (including four carotenes that are the natural precursors of vitamin A), vitamin C, and vitamin E. Micronutrients such as selenium work synergistically with vitamin E to enhance its effectiveness. In addition, selenium and the sulfur-containing amino acids such as cysteine are vital building blocks for the powerful antioxidant enzyme glutathione.

Exactly how do these potent substances help protect you from cancer? The answer lies in the way cancer starts, a step called initiation.

At the root of all cancer is genetic damage. This doesn't mean that cancer is always inherited—researchers estimate that only about 5 percent of all cancers have an inherited basis. Rather, genetic damage is the result of repeated attacks on the genes in the nucleus of your body's cells. If the almost imperceptible accumulation of cellular damage causes just one single cell in your body to lose its normal restraints on growth, it can start to multiply in a disorganized, uncontrolled way. Cancer has been initiated.

The genetic damage that triggers cancer initiation can have many causes. Cigarette smoke, for example, is a powerful carcinogen (a substance that promotes cancer). In fact, many combustion products, such as the hydrocarbons found in smog and air pollution, are carcinogenic. Benzene, a chemical used in many industrial applications, is known to cause leukemia. In all, researchers have identified at least four hundred commonly used chemicals that can cause cancer; thousands more are in common use but have never been tested for their cancer-causing potential. In today's industrial society, we are all constantly exposed to some of these substances.

When carcinogens combine with the DNA in your genes or with any other cellular protein in your body, they form a complex called an adduct. Carefully controlled scientific studies have shown that people with high levels of adducts in their blood are markedly more likely to develop cancer.

Carcinogens and the adducts they form are not the only cause of cellular damage. After all, even nonsmokers sometimes get lung cancer. The free radicals that are constantly released as a normal part of our metabolism are often responsible for cancer-causing cellular damage. Those dangerous oxygen molecules can cause damage to cell membranes and nuclei if they are not mopped up rapidly.

Antioxidants are your body's natural defense against the genetic disruption that initiates cancer. The glutathione and other antioxidant enzymes made by your body neutralize free radicals from environmental poisons and from normal metabolism. Glutathione also captures toxins and heavy metals and removes them from your body. Vitamins A, C, and E function as important antioxidants, as do many other natural substances found in a healthy diet rich in fresh fruits and vegetables. Carotenes, for example, are found in orange, yellow, and red fruits and in vegetables such as collards. The carotenes help protect your cells against cancer not only by acting as powerful antioxidants but by inhibiting the uptake of carcinogens into the cells to begin with.

Study after study has shown that high levels of antioxidant vitamins and enzymes are essential for protecting you against the environmental toxins that surround us every day and for handling the free radicals you normally produce. Low blood levels of antioxidants or vitamins are a clear risk factor for cancer, especially cancer of the lung, esophagus, cervix, and breast. High levels of antioxidants and vitamins have been shown to protect against many kinds of cancer, including melanoma (skin cancer), lung cancer, colon cancer, and bladder cancer—even among people who smoke.

When a cell in your body becomes cancerous, your body deploys its natural defenses to remove the marauder. In most people most of the time, the occasional cancerous cell is dealt with very effectively by the immune system. If your antioxidant and vitamin levels are high, your immune system will function optimally and destroy the renegade cells quickly. If your immune system is compromised (by HIV or AIDS, for example), or if your antioxidant and vitamin levels are inadequate due to low intake or poor absorption, cancerous cells may escape detection and destruction. They may then grow rapidly, forming a mass—a tumor—that eventually can develop its own blood supply and affect your body's functions.

Even after a tumor has formed, high levels of antioxidants and vitamins in your blood could help keep it under control by slowing its growth rate and activating your immune system to fight back. Your infection-fighting white blood cells, for example, function best when they contain plenty of vitamins C and E. Carotenes may help inhibit the genetic damage that can lead to proliferating cancer cells; they may also inhibit the formation of blood vessels that nourish the tumor.

The mineral selenium is another essential component of the anticancer nutrients. Your body needs selenium to produce glutathione. Selenium functions synergistically with vitamin E, making it work more effectively, and is also necessary for good immune function. In a landmark study in the 1980s, researchers looked at dietary selenium intakes among people living in different parts of China. They proved a firm connection between low selenium levels and increased cancer levels. Other studies have also shown that people with low blood levels of selenium and vitamin E are significantly more likely to develop cancer.

REDUCING YOUR RISK

At every stage in the cancer cascade, antioxidants and vitamins can reduce your risk and help prevent the cancer from growing faster or spreading to other parts of your body. I strongly recommend that everyone take the basic supplements discussed in chapter 1.

Simply taking supplements will help reduce your cancer risk, but they are not magical—there are no guarantees. If you take lots of extra vitamins and continue to smoke, for example, you are still at high risk for cancer, to say nothing of emphysema and heart disease. The best approach is to combine antioxidant supplements with other steps for maximum cancer protection.

Try to reduce your exposure to carcinogens. In today's society this isn't easy, but Figure 4.1 lists some simple steps that can reduce your exposure. If you must work in a toxic environment, wear protective clothing and take any other possible preventive steps (improved ventilation, for example). Today's "sealed" office buildings, with windows that can't be opened, often contain construction materials, carpets, and furniture that contain carcinogenic chemicals. The lack of ventilation allows dangerous toxins to build up in the environment. We've already discussed the dangers of sick building syndrome in chapter 3; if you work in such

Figure 4.1

AVOIDING CARCINOGENS

- If you smoke or use any tobacco products, stop. Avoid second-hand smoke whenever possible.
- Limit your exposure to sunlight. Use sunscreen with a high SPF (sun protection factor) when you are outdoors for an extended period.
- Avoid alcoholic beverages.
- Limit your consumption of smoked, pickled, or salted foods. Avoid nitrates in processed foods.
- Minimize exposure to automobile exhaust, gasoline fumes, and air pollution.
- Avoid exposure to chemical fertilizers, pesticides, herbicides, and fungicides.

a building, reread that chapter to learn how to reduce your exposure to possible carcinogens.

Another good way to reduce your exposure to carcinogens is to drink only pure, filtered water. Municipal water systems today routinely add chemicals such as chlorine to your drinking water. Some areas have hard water that contains large amounts of dissolved minerals such as magnesium and iron. Excessive iron in the body is just as bad as too little—among other things, iron catalyzes (speeds up) free radical activity that can lead to cellular damage. Some researchers estimate that more than 10 percent of the American population has iron overload. The pipes in your local water system and your home can add heavy metals such as copper, cadmium, and lead to the water you drink. I strongly recommend that you install a good water filtration system such as Water Pure in your home. If you prefer to use bottled water, purchase only those brands in glass bottles. Water in plastic containers may be contaminated by the chemicals in the plastic.

Exposure to sunlight is a leading cause of skin cancer—and doctors are seeing more and more cases of skin cancer every year. Skin cancer can often be detected early and treated successfully, but it can sometimes lead to malignant melanoma, a very dangerous form of cancer. Use a sunscreen with a high sun protection factor (SPF) when you are outdoors.

DIET AND CANCER PREVENTION

Perhaps the most important step you can take for cancer prevention is to improve your diet. Although the National Cancer Institute points out that there is no evidence that any kind of diet or food can either cure cancer or stop it from returning, anecdotal evidence suggests otherwise. There's also plenty of evidence to show that eating a healthful diet could help prevent the occurrence of the cancer to begin with. Eat plenty of fresh fruits and vegetables, whole grains, and dietary fiber. Limit fats in your diet, and maintain your weight at normal levels.

Fresh fruits and vegetables supply your body with lots of antioxidant vitamins and phytochemicals, along with dietary fiber. Scientists today are very excited by what they are learning about phytochemicals. Sometimes called flavonoids, phytochemicals are substances found in plant foods. The phytochemical lycopene, for example, is a powerful carotene antioxidant that your body does not convert to vitamin A. Lycopene is found in tomatoes, watermelon, and other red-colored foods. Cruciferous vegetables such as broccoli and cabbage contain antioxidant phytochemicals such as sulforaphanes and isothiocyanates. These sulfur-containing substances are important antioxidants in their own right, and they also provide the crucial sulfur molecules your body needs to manufacture glutathione. Other valuable antioxidant phytochemicals include quercetin (found in onions), ajoene (found in garlic), coumarins (found in citrus fruits), indoles (found in cruciferous vegetables), and many others.

Some very exciting recent work with substances found in soybeans may be of great help in preventing and treating cancer. Many edible plants contain isoflavones (also called phytoestrogens), complex, naturally occuring compounds with powerful antioxidant and anticancer effects. Soybeans are very rich in isoflavones, including one called daidzein, and are the only source of an isoflavone called genestein. Careful studies suggest that genestein may help inhibit the growth of cancer cells and the blood vessels that nourish them. In addition, genestein is a powerful antioxidant. Less is known about daidzein, but it too shows great potential in fighting and preventing cancer. Researchers believe these isoflavones are particularly effective against cancers of the colon, prostate, breast, and uterus because these cancers are all sensitive to estrogen, a hormone naturally produced by your body. Estrogen stimulates the cancer cells and makes them grow out of control. Genestein, a phytoestrogen, may work by binding to the estrogen receptors in the cancerous cells; this blocks hormonal estrogen from entering the cells. Because the phytoestrogens are far less powerful than hormonal estrogen, they don't stimulate the cancer cells to grow.

If you are at risk for cancer because of high levels of exposure to carcinogens—if you have recently stopped smoking, for example—it is particularly important for you to maintain high levels of cancer-fighting antioxidants. You also need extra protection if you have a family history of cancer, particularly cancer of the colon or reproductive organs. In addition to the basic supplements recommended in chapter 1, I also suggest you take genestein and daidzein supplements, now available from Solgar and other reliable manufacturers.

It's clear that a diet rich in vitamins and antioxidants can help protect you against cancer. Unfortunately, we can't always eat at least five servings a day of fresh fruits and vegetables. Many of us have such frantic schedules that we just don't have time to prepare healthful meals often enough. It's hard to eat properly when you are dining out or traveling or if you are ill. And even if you do eat a good, healthful diet all the time, it may not be enough to protect you against cancer. Why? Because, as we discussed above, we are all unavoidably bombarded with carcinogenic substances every day. I feel strongly that we all need more antioxidant protection than we can reasonably get through diet alone. For maximum protection, eat well *and* take the supplements recommended in chapter 1. In addition, I recommend taking Phytogreens powder, a broad-spectrum cruciferous vegetable formula. The usual daily dose is half a scoop mixed into six to eight ounces of cool fruit juice.

LIVER FUNCTION AND CANCER

Maintaining good liver function is another significant step you can take for cancer protection. As discussed in chapter 2, your liver plays an essential role in detoxifying your body. You may recall that many substances become dangerous carcinogens only when they are broken down or biotransformed by the family of cytochrome P-450 liver enzymes in the oxidation step, or Phase I detoxification. People vary in their natural ability to produce all the many P-450 enzymes. Minor variations in how well your liver enzymes biotransform chemical wastes can play a major role in your chances of getting cancer. One type of P-450 enzyme called CYP1A1, for example, biotransforms polycyclic aromatic hydrocarbons (PAHs), a common carcinogen found in combustion products such as cigarette smoke. In some people, the CYP1A1 enzyme actually activates the PAHs into a more carcinogenic form. This could explain why only about one out of ten cigarette smokers gets lung cancer—the ones who do are probably the ones with the defective enzyme. Other individual variations among the cytochrome P-450 enzymes can affect how well your liver deals with other carcinogens.

In the conjugation step of detoxification, Phase II, glutathione and other enzymes in your liver corral free radicals, heavy metals, and some toxic wastes (including PAHs) and remove them from your system. Individuals vary in their production of these enzymes as well, making them more vulnerable to some types of cancer, particularly cancer of the bladder and lung.

You can't change your individual genetic makeup, but you can do everything possible to help your liver deal with waste products efficiently and compensate for normal variations in your liver enzymes. It is particularly important to maintain a high level of glutathione. I recommend taking vitamin and NAC supplements and GSH 250 Master Formula or Glutaplex (see chapter 1) to maintain a healthy liver. Speak to your health care provider about tests for liver function, particularly the functional liver detoxification profile covered in chapter 2. If the tests indicate your liver function is below par, consider a program of metabolic cleansing. Even if your liver function is normal, I recommend following a short metabolic cleansing program at least once a year, and more often if you are frequently exposed to carcinogens. If you work in a sealed building, in a garage, or in any other environment that contains chemical toxins (a print or copy shop, for example), discuss more frequent metabolic cleansing with your health care provider.

Some studies suggest that stress can play a role in causing cancer. The hormones your body produces in response to stress are broken down in your liver through a process called sulfation. If you have an overload of stress hormones, your liver will have to work extra hard to deal with them. Because sulfation needs lots of the sulfur-containing amino acid cysteine to occur, when you're under stress your liver diverts cysteine to sulfation. Unfortunately, that means there's not much cysteine left over to make glutathione—which in turn means that some dangerous biotransformed wastes and free radicals escape capture and removal. Instead, they continue to circulate in your body and damage your cells. It's no surprise that a major study in Sweden over a ten-year period showed an increase in colon and rectal cancer among people who had been under serious stress, such as prolonged unemployment or divorce.

If you are often stressed from work or your personal life, your liver needs some extra assistance to be sure it detoxifies carcinogens and other wastes along with the stress hormones. See page 25 in chapter 2 for my supplement recommendations. Talk to your health care provider about doing a metabolic cleansing program.

TREATING CANCER

Once cancer has been detected in your body, amino acids and antioxidants can be valuable tools in treatment to regain your health. Cancer treatment varies considerably depending on the type of cancer and the stage it is in. In general, treatment involves surgery, chemotherapy, and radiation therapy, alone or in combination. Half of all cancer patients are treated with radiation. In fact, for many patients, radiation is the only treatment needed. Radiation is sometimes used to shrink tumors before they are removed surgically. In other cases, radiation is used to relieve cancer symptoms. Chemotherapy uses powerful drugs to treat cancer. Anticancer drugs destroy cancer cells by stopping them from growing or multiplying at one or more stages in their life cycle. Because some drugs work better together than alone, chemotherapy often involves more than one drug. Chemotherapy is also sometimes combined with surgery or radiation treatments. Depending on the type of cancer and its stage of development, chemotherapy can be used to cure the cancer, keep it from spreading, slow its growth, kill cells that have spread (metastasized) to other parts of the body, and relieve symptoms.

Chemotherapy and radiation treatment usually have side effects such as nausea, vomiting, hair loss, and fatigue. Here's where amino acids and antioxidants can be very helpful. Cysteine in the form of NAC has been shown to help reduce the severe side effects of some chemotherapy drugs and radiation treatment. Glutathione can also help protect against the side effects of radiation treatment. This seems to work best if your glutathione level is high before the treatment begins. If you are preparing to undergo radiation therapy for cancer, discuss glutathione supplements with your doctor before you start.

If you ordinarily take glutathione or cysteine supplements or large doses of antioxidant vitamins, be sure to tell your doctor before treatment starts. If you wish to start taking these supplements as part of your cancer treatment, discuss the question with your doctor first. In some types of cancer, a rise in your blood level of glutathione is an indication that the cancer treatment—either chemotherapy or radiation—is working. In that case, taking glutathione or cysteine supplements could cause a misleading rise in your glutathione level.

Chemotherapy and radiation therapy cause very large amounts of free radicals to be released in your body. If the free radicals aren't counteracted, they will do damage to your cells precisely at a time when they are very vulnerable. During cancer treatment it is extremely important that you take in enough antioxidant vitamins and nutrients to counteract the increased free radicals. Numerous studies have shown that patients who take supplements during cancer treatment often have fewer or less severe side effects. Some very exciting recent studies suggest that antioxidants in large doses can be almost as helpful as the cancer drugs and radiation for reducing tumors. More important, the studies suggest that patients who get sufficient antioxidant support during cancer treatment are less likely to have recurrences or secondary cancers later on. The doses needed are large and should be taken only under medical supervision.

Weight loss, lack of appetite, and lack of energy from cancer treatment are unavoidable to some extent, but amino acids can help keep your weight and energy levels up. Free-form mixed amino acids, combined with pure filtered water or diluted fruit juice, are easy to prepare and swallow. They provide the building-block proteins your body needs to keep up your strength. The amino acid glutamine is particularly valuable for helping your small intestine recover and for preventing muscle wasting. I usually recommend 3 to 4 g a day of free-form amino acids, plus an additional 2 to 3 g of glutamine. (Too much glutamine may cause mild constipation; if this occurs, try reducing the dose or adding more soluble fiber to your diet.) In addition to the antioxidant vitamins, be sure to get enough of the other vitamins and minerals by taking a high-potency, no-iron formula from a reputable manufacturer such as Solgar, TwinLabs, or Nature's Plus.

Cancer is a difficult illness to treat. Every case is different and must be handled individually. If you must undergo cancer treatment, discuss nutrition and supplemental antioxidants with your doctor *before* you start. If your doctor seems to resist the idea, I urge you to find another doctor who is more open-minded. (See Figure 4.2 for help in finding nutritionally oriented or alternative cancer treatment.)

Figure 4.2

SOURCES OF CANCER INFORMATION AND TREATMENT

If you have been diagnosed with cancer, you may face a confusing array of treatment options and a barrage of unfamiliar medical terminology. The organizations listed below can help you understand some of the basic medical issues.

SOURCES OF CANCER INFORMATION AND TREATMENT (cont.)

American Cancer Society
1599 Clifton Road NE
Atlanta, GA 30328
(404) 329-7647

National Cancer Institute
Office of Cancer Communications
9000 Rockville Pike
Bethesda, MD 20892
(800) 4-CANCER

For information on finding nutritionally oriented cancer treatment, contact:

American Institute for Cancer Research
1759 R Street NW
Washington, DC 20009
(800) 843-8114

Arlin J. Brown Information Center, Inc.
Box 251
Fort Belvoir, VA 22060
(703) 752-9511

Foundation for the Advancement of Innovative Medicine (FAIM)
Two Executive Boulevard, Suite 204
Suffern, NY 10901
(914) 368-9797

CURING CANCER

Experimental work with rats in the 1980s suggests that glutathione could actually reduce or even cure some kinds of cancer. The rats were fed a potent carcinogen that causes liver tumors. After the tumors developed, some of the rats were treated with intravenous glutathione while others received no treatment. Among the treated rats, 80 percent lived, while all the untreated rats died. The liver tumors among the rats that survived had shrunk considerably or even disappeared.

Does this mean that glutathione can cure cancer in humans? No. Although more and more researchers believe that glutathione has a role in cancer prevention and treatment, there is no single substance that "cures" cancer. Today and for the foreseeable future, prevention through careful attention to diet, antioxidants, and reduced exposure to carcinogens is the best approach.

REFERENCES

Bendich, Adrianne, E. Gabriel, and L. J. Machlin. "Effect of Dietary Level of Vitamin E on the Immune System," *Journal of Nutrition*, vol. 113, pp. 1920–26, 1983.

Block, Gladys. "Dietary Guidelines and the Results of Food Consumption Surveys," *American Journal of Clinical Nutrition*, vol. 53 (supplement), pp. 356S–57S, 1991.

Block, Gladys. "Vitamin C and Cancer Prevention: The Epidemiological Evidence," *American Journal of Clinical Nutrition*, vol. 53 (supplement), pp. 270S–82S, 1991.

Bowen, R., et al. "Antipromotional Effect of Soybean Isoflavone Genestein," *Proceedings of the American Association for Cancer Research*, vol. 34, p. 555, 1993.

Braakhuis, J. M. Boudewijn. "Antioxidant-Related Parameters in Patients Treated for Cancer Chemoprevention with N-acetylcysteine," *European Journal of Cancer*, vol. 31A, no. 6, pp. 921–23, 1995.

Byers, Tim, and Geraldine Perry. "Dietary Carotenes, Vitamin C, and Vitamin E as Protective Antioxidants in Human Cancers," *Annual Review of Nutrition*, vol. 12, pp. 139–59, 1992.

Coward, L., et al. "Genestein, Daidzein, and Their B-glycoside Conjugates: Antitumor and Isoflavones in Soybean Foods from American and Asian Diets," *Journal of Agriculture and Food Chemistry*, vol. 41, pp. 1961–67, 1993.

De Flora, S., et al. "Antioxidant Activity and Other Mechanisms of Thiols Involved in Chemoprevention of Mutation and Cancer," *The American Journal of Medicine*, vol. 91 (supplement 3C), September 30, 1991.

De Vries, N., and S. De Flora. "*N*-acetyl-L-cysteine," *Journal of Cell Biochemistry*, vol. 17F (supplement), pp. 270–77, 1993.

Giovannucci, Edward, et al. "Folate, Methionine, and Alcohol Intake and Risk of Colorectal Adenoma," *Journal of the National Cancer Institute*, vol. 85, no. 11, pp. 875–81, 1993.

Hughes, George S. "The Effects of Beta-Carotene on the Immune System in Cancer," *Nutrition Reports*, vol. 10, no. 1, pp. 1–8, 1992.

Jaakola, K., et al. "Treatment with Antioxidant and Other Nutrients in Combination with Chemotherapy and Irradiation in Patients with Small-Cell Lung Cancer," *AnticancerResearch,* vol. 12, pp. 599–606, 1992.

MacNee, William, et al. "The Effects of N-acetylcysteine and Glutathione on Smoke-Induced Changes in Lung Phagocytes and Epithelial Cells," *The American Journal of Medicine,* vol. 91 (supplement 3C), September 30, 1991.

Malone, W. F. "Studies Evaluating Antioxidants and Beta-Carotene as Chemopreventives," *American Journal of Clinical Nutrition,* vol. 53 (supplement), pp. 305S–13S, 1991.

Martin, P. M., et al. "Phytoestrogen Interaction with Estrogen Receptors in Human Breast Cancer Cells," *Journal of Endocrinology,* vol. 103, pp. 1860–67, 1978.

Messina, M., et al. "Soy Intake and Cancer Risk: A Review of the In Vitro and In Vivo Data," *Nutrition and Cancer,* vol. 21, pp. 113–31, 1994.

Novi, A. A., R. Florke, and M. Stukenkemper. "Regression of Aflatoxin B_1-Induced Hepatocellular Carcinomas by Reduced Glutathione," *Science,* vol. 2121, no. 5, pp. 541–42, 1981.

Passwater, Richard A. *Cancer Prevention and Nutritional Therapies* (New Canaan, Conn.: Keats Publishing, 1993).

Passwater, Richard A. *The New Supernutrition* (New York: Pocket Books, 1991).

Perera, Frederica P. "Uncovering New Clues to Cancer Risk," *Scientific American,* vol. 274, no. 5, pp. 54–62, May 1996.

Singh, V. N., and S. K. Gaby. "Premalignant Lesions: Role of Antioxidant Vitamins and Beta-Carotene in Risk Reduction and Prevention of Malignant Transformation," *American Journal of Clinical Nutrition,* vol. 53 (supplement), pp. 386S–90S, 1991.

Weisburger, J. H. "Nutritional Approach to Cancer Prevention with Emphasis on Vitamins, Antioxidants, and Carotenoids," *American Journal of Clinical Nutrition,* vol. 53 (supplement), pp. 226S–37S, 1991.

Yu, Shu-Yu, et al. "Regional Variation of Cancer Mortality Incidence and Its Relation to Selenium Levels in China," *Biological Trace Element Research,* vol. 7, pp. 21–29, 1985.

GLUTATHIONE
AND AGING

Some 12 percent of the American population today—about thirty million people—is over the age of sixty-five. That large number will jump sharply upward in the next fifteen to twenty years, as the huge baby boom generation born between 1946 and 1964 reaches retirement age. Anyone who is aged sixty today has a very good chance of living another fifteen to twenty years or more.

Most older Americans eventually suffer from heart disease, stroke, Alzheimer's disease, arthritis, cancer (see chapter 2), and other debilitating ailments. Yet many have somehow escaped these problems or have them to only a limited degree. They have easily achieved the age of eighty or older in vigorous good health, both physical and mental. How do they do it? How can you?

The answers are complex and by no means definite. Gerontology (the study of aging) is a new science, one that so far has provided many suggestive ideas and tantalizing clues as well as some well-grounded theories as to why we age. Lifestyle choices may play an important role. Studies of centenarians in the United States show that very few ever smoked and that most drank alcohol sparingly or not at all. Other lifestyle factors that may be related to longevity include eating a low-fat diet and staying physically active.

There's probably no one single cause of aging, but many researchers believe that cellular damage from free radicals may play an important role. It's quite possible that a reduction in the number of free radicals in the body delays the onset of cancer, heart disease, degenerative nervous system diseases, and loss of immune function. High levels of antioxidants may also keep damage to the genetic apparatus in the cell to a minimum, which could reduce your likelihood of disease. People with high levels of antioxidants in their bodies to mop up destructive free radicals may therefore have a longer and healthier life expectancy. In particular, a high level of glutathione may make the difference. Humans have a much higher glutathione level than do other mammals, and humans also have much longer lives. In a 1994 study of a cross-section of elderly people in Southfield, Michigan, those with the highest glutathione level were very healthy—and also very old, ranging from eighty to ninety-five.

Among my older patients, I have found a similar correlation. Charlie, an eighty-four-year-old retired police officer, came to me because he just didn't have his old stamina and sharpness. When he sat down to read or watch TV, he fell asleep instead. He was getting more and more forgetful. His doctor said he couldn't find anything wrong and that his symptoms were normal for a man his age. Charlie refused to accept this diagnosis and came to see me instead.

When we discussed his diet, Charlie said he ate mostly fruit, grains, and salads. He ate no animal protein—no milk, no cheese, no eggs, no fish, no meat. He didn't exercise and he took no supplements.

After talking with me, Charlie decided to add a small daily serving of fish, chicken, lamb, or eggs to his diet. I suggested a digestive enzyme supplement to help his system adjust to the change. I also suggested a daily multivitamin with minerals and three capsules daily of GSH 250 Master Formula supplement. I told him it could take a month or six weeks for the positive effects of glutathione and more amino acids to be felt.

To my surprise, Charlie called me just five days later. He felt great! His energy levels were up markedly, his memory was improved, and he was walking a mile a day. Five years later, he's still going strong.

There's no way of knowing for certain what the future holds for your health. Even so, you can make some nutrition and lifestyle choices now that will have a positive impact on your longevity.

HEART DISEASE

The statistics for heart disease among Americans are truly frightening. In 1995 alone, about one million Americans died of heart disease; about half of those deaths were from heart attacks directly related to coronary artery disease. In addition, thousands more underwent bypass surgery and other procedures. The total cost of treatment for heart disease runs to about fifty billion dollars *a year*.

At the root of most heart problems is coronary artery disease caused by atherosclerosis. Your heart muscle is nourished by blood carried to it by two coronary arteries that branch off from your aorta, the main artery carrying blood from your heart. For a variety of complex reasons that we'll discuss, fatty deposits called atheromas can form on the inner walls of those arteries—a condition called atherosclerosis. The deposits narrow the artery and make it less flexible, which keeps your heart from getting enough oxygen-rich blood. If the artery becomes so clogged that it is completely blocked or goes into a spasm, or if a blood clot (thrombus) forms and blocks the artery, you will have a heart attack (myocardial infarction).

Some people with blocked coronary arteries develop angina. When they exert themselves, their stiff, narrowed coronary arteries can't expand enough to increase the blood flow and give the heart muscle more oxygen. The result is severe chest pain; in serious cases, angina can be very disabling.

Atherosclerosis can also develop in other major arteries. If the carotid artery in the neck or an artery in your brain is blocked, you could have a stroke. If the arteries leading to the kidneys are blocked, you could develop kidney failure. If the blockage is in the arteries leading to your legs, the poor circulation could lead to fatigue, muscle cramps, or even gangrene.

Because we often use words such as *blocked* or *clogged* when discussing atherosclerosis, we tend to think of it as a sudden development, like a clogged drain. In fact, a much better plumbing analogy is of lime deposits gradually building up in a water pipe. Atherosclerosis is caused by a complex cascade of events over a long period of time. The fatty deposits in your arteries may have started to form when you were in your twenties. The entire process of atherosclerosis is slow, insidious, and often without symptoms. Angina or a heart attack could be your first—and maybe your last—warning.

Here's how the process begins. Your arteries actually have three layers. On the outside is the adventitia, made of tough connective tissue. The middle layer is the media, made of smooth muscle cells that can contract or dilate, thus lowering or raising your blood pressure. The inner layer is called the endothelium. It's made of very smooth, squarish intima cells that look a lot like tiles on a floor. The tubular part of the artery through which the blood flows is called the lumen. Atherosclerosis begins when the intima cells that form the smooth inner lining of the artery are damaged. As a result of the injury, the intima cells are disrupted and the lining of the artery is no longer perfectly smooth. Your body naturally sends a fatty substance called cholesterol to the area in an attempt to repair the damage. When fat is exposed to oxygen, however, it oxidizes—put more simply, it goes rancid. Not only does this oxidation create damaging free radicals, the oxidized cholesterol itself then attracts a type of immune cell called a monocyte. The monocytes attach themselves to the cholesterol and manage to get under the intima cells and into the media, the center layer of the artery. Once inside the media, the monocytes oxidize the cholesterol even more and form foam cells. At first, the cholesterol-laden foam cells form fatty streaks inside the artery. As the process continues, smooth muscle cells in the media begin to form around and under the cholesterol deposits, pushing them out into the lumen of the artery and forming an atheroma. The atheroma attracts additional cholesterol and grows larger as additional fatty deposits and foam cells build up. Calcium from your blood is also attracted to the atheroma, which helps make the fatty deposit harden into plaque. Adrenaline, a natural hormone in your blood, roughens the surface of the atheroma. For reasons scientists don't quite understand yet, this in turn attracts fibrinogen from your blood. Fibrinogen is part of your body's natural repair system for forming clots to stop bleeding. It makes a sort of fibrous network over the fatty deposit. The accumulated calcium and fibrinogen gradually make the atheroma fibrous and tough. The fibrinogen in the atheroma may attract platelets, the blood cells that aid in clotting. Eventually, a blood clot could form on top of the atheroma, blocking the artery still further. If the clot blocks the artery completely or breaks off and lodges further along in the artery, you could have a heart attack.

Cholesterol, Antioxidants, and Your Heart

Cholesterol is involved almost from the start of the atheroma cascade, but what is it exactly? Cholesterol is a waxy, fatty substance your body manufactures in your liver from dietary fat. It's essential for making and repairing cell membranes and the sheaths that cover your nerves. It's also vital for metabolizing vitamin D and making many of the hormones that regulate your body. Although your liver makes all the cholesterol you need from the foods you eat, you also get dietary cholesterol from foods such as eggs, meat, fish, milk, and other dairy products. Plant foods have no cholesterol, which is why many unhealthy snack foods fried in vegetable oil can nonetheless proclaim that they are cholesterol-free.

Because cholesterol is a fat, it doesn't mix with water—and your blood is mostly water. To get the cholesterol (as well as fatty acids and triglycerides) from your liver to where it needs to go in your body, tiny droplets of cholesterol are enclosed in a sphere of protein. Some of the spheres are made of low-density lipoprotein (LDL). The LDLs transport cholesterol to the cells. Your liver also makes high-density lipoprotein (HDL). These spheres collect the unused LDLs and return them to the liver, where they are removed from the bloodstream and excreted through the bile.

Cholesterol is often regarded with horror as the source of all arterial evil. Many people mistakenly believe that all low-density lipoproteins are intrinsically bad and that the lower their LDL level is, the less likely they will be to have heart disease. This isn't quite true. A better measure of your heart disease risk is the ratio of LDL to HDL. For example, if your LDL level is high but your HDL level is also high, it offsets some of the risk. Two standard ratios help determine if your cholesterol levels are within healthy ranges. To determine the first, divide your total cholesterol level by your HDL level. Ideally, the result will be below 4. To determine the second benchmark, divide your LDL level by your HDL level. Ideally, the result will be below 3.

Elevated LDL levels are associated with a higher risk of heart disease, but it's not the LDLs themselves that are the problem—it's the free radicals that are created when the LDLs are oxidized. However, if your LDL level is within the normal range, you will have less cholesterol to oxidize and will therefore produce fewer damaging free radicals.

As you grow older, it's important to know your cholesterol levels and try to keep them within the generally accepted healthy range (see Figure 5.1). Any blood cholesterol level of 200 milligrams per deciliter (mg/dL) or more increases your risk of heart disease. A level of 240 mg/dL or greater is considered high cholesterol. If you have this level, you are at more than twice the risk of heart disease of someone with desirable cholesterol. Unlike total cholesterol, where a lower number is better, the higher your HDL ("good") cholesterol number, the better. If you do have high cholesterol, studies show that lowering it even a little can have a positive effect on your health. A healthy forty-year-old man with borderline high cholesterol (200 to 240 mg/dL) who lowers his cholesterol by just 10 percent can cut his risk of a heart attack by 50 percent.

If your LDL cholesterol is too high, your doctor may suggest that you attempt to lower it through diet by restricting your intake of fats. If this doesn't help—and it often doesn't—the standard medical approach is often to prescribe powerful (and possibly dangerous) cholesterol-lowering drugs such as lovastatin (Mevacor) or probucol. These drugs have some significant drawbacks, however. Among other problems, they can inhibit your production of co-enzyme Q-10, an important heart nutrient.

If your cholesterol levels are very high, if you have been diagnosed with serious atherosclerosis, if you have angina, or if you have had a heart attack, I suggest you discuss a very low-fat dietary regimen called the Ornish program with your physician. This program was developed by Dr. Dean Ornish of the University of California, San Francisco, School of Medicine. When combined with stress reduction training and exercise, the program can help reduce cholesterol levels and may even reverse plaque buildup in your arteries. While the Ornish program can be very helpful for some people with severe heart disease, it is controversial. New ev-

Figure 5.1

TOTAL BLOOD CHOLESTEROL/HDL CHOLESTEROL

Level*	Category
less than 150 mg/dL	undesirable
less than 200 mg/dL	desirable
200 to 239 mg/dL	borderline high
240 mg/dL	high
HDL less than 35 mg/dL	low

LDL CHOLESTEROL

Level	Category
less than 100 mg/dL	undesirable
less than 130 mg/dL	desirable
130 to 159 mg/dL	borderline high risk
160 mg/dL and above	high risk

*Note: Blood cholesterol levels are measured in milligrams per deciliter, abbreviated as mg/dL.

idence suggests that your body needs more protein than the Ornish program permits.

If your cholesterol levels are high but not dangerously so, and if you have little or no evidence of significant coronary artery disease, I urge you to talk to your doctor about lowering your cholesterol through natural means before you go on a very restricted diet or start taking potentially harmful or unnecessary drugs. I know from working with my patients that changing your diet and adding antioxidant supplements can be very effective. (See the sidebar on page 113 for information about some exciting new natural treatments for high cholesterol.)

A low-fat, high-fiber diet with little meat, little sugar, and plenty of fresh fruits and vegetables is the first step. Use only monounsaturated oils such as olive or canola oil. Garlic and onions help thin the blood and reduce the "stickiness" of platelets, which can lower the chances of a blood clot. Omega-3 fatty acids in fish oil or flaxseed oil can also help reduce platelet stickiness.

Avoiding eggs has become almost a phobia among people with

high cholesterol, but some very convincing studies indicate that eating eggs every day can actually raise your HDL levels. As we've discussed, it is just as important to raise your HDL ("good") cholesterol as it is to lower you LDL ("bad") cholesterol. One egg yolk contains about 200 mg of dietary cholesterol and about 5 g of fat, including only 2 g of saturated fat. (Egg whites have no fat or cholesterol.) Eggs are an excellent source of vital amino acid building blocks, iron, and other nutrients. Although most physicians today suggest you limit your intake to just four egg yolks a week, I usually recommend that my patients eat one egg a day. (See chapter 9 for a more extensive discussion of eggs.)

The next step is to add antioxidant supplements. Vitamins E and C are powerful antioxidants that can help keep free radicals from damaging your arteries; along with the mineral magnesium, they also help reduce the stickiness of your platelets, which helps prevent dangerous blood clots. It's vital to maintain high levels of these essential nutrients as you grow older. You also need to maintain high levels of other vitamins and of vitamin cofactors such as selenium. If you don't get enough vitamin C, for instance, you are losing more than the protection it gives you against free radicals. Vitamin C is also essential for maintaining the connective tissue that holds your arteries together. A shortage of vitamin C can weaken the arteries and let blood leak into the walls, starting the atheroma cascade. The mineral selenium works synergistically with vitamin E; an inadequate level of either nutrient means the other will operate less efficiently and provide less protection. Inadequate amounts of vitamin B_6 (pyridoxine) can also cause arterial leakage. In addition, vitamin B_6 is essential for metabolizing the sulfur-containing amino acids cysteine and methionine. Without sufficient vitamin B_6, homocysteine, a byproduct of cysteine and methionine metabolism, can build up in your bloodstream and cause atherosclerosis.

Carnitine is another very valuable amino acid for keeping your heart healthy. This nonessential amino acid can help lower blood triglycerides and increase your level of HDL ("good") cholesterol. It can also help relieve angina symptoms and heart arrhythmias. Carnitine is found chiefly in muscle meat such as beef or pork. If you eat little or no meat as part of your cholesterol-lowering diet, it's particularly important to take the supplement cysteine, in the form of NAC, which can help lower blood cholesterol and raise your glutathione level; lipoic acid will increase NAC's effectiveness.

My patient Michael is a good example of how diet and anti-oxidant supplements can help reduce high cholesterol. Michael, a fifty-three-year-old psychology professor, had a family history of early heart disease. Although he ate what he thought was a healthy diet for someone with a potentially bad heart, his total cholesterol level was 250. His HDL level was very low, while his LDL was on the high side. Michael had no symptoms and exercised regularly, but he quite rightly feared he was a good candidate for a heart attack. His doctor wanted to put him on cholesterol-lowering medication. Rather than take powerful drugs, Michael came to me to learn ways to lower his cholesterol through diet and supplements.

Michael's heart-healthy diet contained less than 10 percent protein. He ate lots of complex carbohydrates like rice and pasta, which raised his insulin levels. This in turn raised his overall cholesterol, especially his LDL level.

By adjusting his diet to include more protein and fewer carbohydrates, and by adding vitamin E, other antioxidant vitamins, and glutathione, Michael was able to lower his overall cholesterol level and raise his HDL level. The process was gradual, over several months, but today Michael's cholesterol is under control and he no longer fears a heart attack.

Daily Supplements for High Cholesterol

If you have high cholesterol, I suggest you change to a low-fat diet and try these daily supplements for three months. For the best effectiveness, spread the doses out over the day.

Vitamin C	3,000 mg
Vitamin E	400 to 800 IU
Vitamin B_3 (niacin)	50 to 100 mg
Vitamin B_6 (pyridoxine)	25 to 50 mg
Folic acid	400 mcg
Mixed carotenes	25,000 IU
NAC	1,500 mg
Lipoic acid	400 mg

Carnitine	1,000 mg
Magnesium	400 mg
Selenium	250 mcg
Potassium	99 mg
Omega-3 fatty acid	three 100 mg capsules
Garlic	three 500 mg capsules
Coenzyme Q-10	150 mg
Taurine	500 mg
Arginine	1,000 mg

Ask your doctor to check your cholesterol levels again after taking the supplements for three months. In my experience, there's a good chance that you will see a real improvement—your cholesterol levels could drop by 10 percent or more.

If you haven't shown any improvement, it's possible that a sluggish liver is part of the problem. As we discussed above, unused LDL cholesterol is carried to your liver by the HDL cholesterol for elimination from the body. If your liver isn't working efficiently, however, it will not remove the LDL well. Instead, the LDL may continue to circulate in your blood. If you suspect your liver is sluggish, see the recommendations in chapter 2. Improving your liver function may well have a beneficial effect on your cholesterol levels.

Low Cholesterol

If high cholesterol is bad, is low cholesterol good? Not necessarily. Remember, your body needs cholesterol. It is an essential component in many of the vital hormones that regulate your metabolism, and it is essential to your body's ability to repair itself quickly. If your cholesterol levels are within the normal range, don't attempt to lower them by diet or supplements—you'll make your health worse, not better.

Because most of the cholesterol in your body is produced by your liver, low cholesterol levels can be an indication of a sluggish liver. A metabolic cleansing program as discussed in chapter 2 could help. It's also possible that your low cholesterol levels are the result of malabsorption (dysbiosis) in the small intestine. A yeast overgrowth may be affecting your ability to absorb nutrients through your small intestine. See chapter 2 for more information on treating dysbiosis.

Hypocholesterolemia, the medical term for an abnormally low level of cholesterol in the blood, can be a symptom of serious illness such as cancer or liver disease. Medical attention is needed to discover and treat the underlying cause of this condition.

Avoiding Arterial Damage

Damage to the intima cells lining your arteries is the root cause of atherosclerosis. But minor injuries to your arteries occur all the time; your body usually deals with them without incident. Why do some injuries lead to atherosclerosis? Let's look more closely at how the damage starts and how it can be prevented. Atheromas often occur at points in the arteries where the blood flow is naturally disrupted—where arteries branch, for example. The turbulence that occurs at the branch could be the cause of injury. Untreated high blood pressure is another possible cause. The high pressure could cause minute cracks in the inner lining—one of several good reasons for keeping your blood pressure under control.

Today, many researchers believe that a significant cause of arterial injury is damage from free radicals and environmental pollutants. And as we've seen, the best way to protect yourself is to maintain high levels of antioxidants in your body. In particular, keep your GSH level high. Other antioxidants such as vitamin E can help deal with free radicals, but you need GSH to corral those dangerous environmental toxins. Heavy metals such as lead have been implicated in arterial damage. Other environmental hazards that could harm your arteries include copper, exhaust fumes, hydrocarbons, chlorine, and many others. Of course, cigarette smoke is packed with toxins.

Antioxidants and Heart Attacks

When the flow of blood through the coronary arteries to your heart is suddenly blocked, the heart muscle is starved of oxygen. The cells of the heart muscle are seriously damaged and cease to work efficiently. In short, you have a heart attack.

Some recent research suggests that the damage from a heart

attack can be limited by immediately administering antioxidants, particularly GSH. The reason is that during the ischemia stage, when the heart cells aren't getting any oxygen, they build up free radicals. The GSH in the cells is used up trying to cope with the metabolic wastes, but no new GSH is available because there is no blood. When the blood flow is restored (reperfusion) and oxygen gets to the cells again, they begin functioning again. Unfortunately, the cells contain free radicals left over from the ischemic period; during reperfusion, they start making even more free radicals. At this point, however, the blood level of glutathione is very low—too low to mop up the proliferating free radicals. Additional cellular damage is the result. It's possible that administering glutathione at this critical point could keep the damage from a heart attack to a minimum.

Some heart attack patients are at risk for dangerous heart arrhythmias that make the heart beat irregularly or cause it to flutter rapidly (fibrillation). Some studies suggest that carnitine can help prevent arrhythmias and fibrillation after a heart attack. Taurine, another essential amino acid, may also play an important role in protecting the heart after a heart attack. Studies show that after ischemia, heart levels of taurine drop sharply. Administering taurine to heart attack patients may help them recover more quickly and completely and avoid some kinds of life-threatening arrhythmias.

GLUTATHIONE AND YOUR BRAIN

Much of what we've discussed regarding atherosclerosis and your heart applies to your brain as well. If the carotid artery in your neck or one of the smaller arteries nourishing the brain is narrowed from atherosclerosis, you may develop cerebrovascular disease. Just as a blockage in a coronary artery leads to a heart attack, a blockage in an artery in the brain causes a stroke.

The same supplements that help lower blood cholesterol and prevent the build up of fatty deposits in the coronary arteries can help prevent it in the brain arteries. To help reduce your risk of stroke, then, I suggest you take the supplements listed on pages 101–102. In addition, I recommend taking ginkgo biloba, a powerful herbal antioxidant that is particularly helpful for the brain. Ginkgo has been shown to help improve the blood flow to the brain, particularly in older people whose blood vessels have been narrowed and hardened by atherosclerosis. It also helps improve the utilization of oxygen and glucose in the brain. This means that even though your brain may be getting less blood—and therefore less oxygen and glucose—ginkgo can lessen the effects. Ginkgo also reduces platelet stickiness, which could reduce your chances of a stroke from a blood clot.

Ginkgo is made from the leaves of the same ginkgo biloba tree (also sometimes called a maidenhair tree) that is a popular ornament on city streets. These trees are botanically very ancient, dating back perhaps 250 million years or more. They are native to southeastern China, where the leaves have been used medicinally for centuries. The active ingredients are found in the fan-shaped leaves. Ginkgo is generally available as a liquid extract that contains about 20 to 25 percent active ingredients, or in capsules containing the dried leaf. I generally suggest taking 40 to 50 mg three times a day. Natrol is a high-quality brand offering both the liquid extract and capsules. The beneficial effects of ginkgo could take up to twelve weeks to be apparent—be patient. If you find ginkgo helps you, continue to take it.

If you are at risk for stroke because of family history, severe high blood pressure, obesity, diabetes, a history of transient ischemic attacks (TIAs, or mini-strokes), or some other reason, discuss using supplements and ginkgo biloba with your doctor before you try them.

Stroke victims whose glutathione levels are low have a poorer prognosis than those whose glutathione levels are higher. This suggests that patients recovering from a stroke would benefit from taking supplements to raise their GSH levels. Researchers are now exploring this interesting avenue of stroke treatment.

A diet that includes eggs every day may be helpful during recovery from a stroke. The nerve cells that make up most of your brain can't store cholesterol, but cholesterol is vital for repairing damaged nerves. Eggs may help provide the extra cholesterol needed to aid in recovery. (We'll discuss eggs more in chapter 9.)

Memory Loss and Forgetfulness

As we age, a certain amount of memory loss is almost inevitable, even if we are otherwise very healthy. This is generally a normal, though annoying, aspect of aging. In normal memory loss, you might forget part of an experience—you might remember everything you had at dinner the night before except for what you had for dessert, for example. You'll probably remember later, however, especially if someone prompts you, and you'll readily admit that you couldn't remember. The good news is that minor forgetfulness is generally just that—minor. Your basic skills remain intact, you can use notes and reminders to compensate for the forgetfulness, and you can still follow directions and care for yourself. In most cases, memory loss is not a first sign of Alzheimer's disease (see below).

Minor as it may be, memory loss is still distressing. It's possible that supplements can help reduce forgetfulness or keep it from getting worse. I think it's particularly important to get adequate levels of the B vitamins, as well as the amino acid glutamine, which are needed to manufacture the neurotransmitter chemicals. Ginkgo biloba can be very helpful for memory loss. Some interesting new research suggests that supplements of phosphatidyl serine (PS), a phospholipid essential for normal brain function, could help improve short-term memory problems. A normal diet contains very little PS, so supplements are the only way to boost your levels. Once very difficult to make and expensive to purchase, PS supplements are now being derived from plant sources and are available from Allergy Research Group. If you'd like to try PS, I suggest 200 to 300 mg daily along with your regular supplements.

If you are troubled by memory loss, I suggest you take an additional 100 mg of vitamin B complex in addition to your daily supplements as discussed in chapter 1. Make sure you get adequate zinc—about 30 mg daily from any high-quality supplement. I also suggest adding 1,500 mg of glutamine and up to 150 mg of ginkgo biloba, divided into three doses over the day. Glutamine is available in capsules from Primary Nutraceuticals.

Alzheimer's Disease

A progressive, degenerative disease that attacks the brain and results in impaired memory, thinking, and behavior, Alzheimer's disease (AD) is the most common cause of dementia in older people. Some four million people in America—including former president Ronald Reagan—now suffer from AD. Because the population is aging, an estimated fourteen million Americans will have AD by the year 2050. At least one in twenty adults over age sixty-five has Alzheimer's; one study estimated that nearly half of all those over age eighty-five have it.

Alzheimer's patients have reduced amounts of neurotransmitter chemicals such as acetylcholine, which are vital for relaying complex messages among the nerve cells of the brain. Autopsies of deceased AD patients show that their brains contained abnormal deposits (plaques) and tangled bundles of nerve fibers.

Free radicals, chemical toxins, and a shortage of antioxidants may be behind Alzheimer's disease. The typical brain plaques and tangles of AD contain calcium. They also contain aluminum, so it is possible that exposure to heavy metals may be one cause of the disease. Because glutathione is a very effective scavenger of heavy metals and other toxins, maintaining a high GSH level may help protect you against Alzheimer's disease. Free radicals can damage the delicate membranes of brain cells; this too may play a role in creating plaques and tangles. Again, maintaining a high level of antioxidants, including GSH, may help protect you.

Many AD patients are deficient in B-complex vitamins, particularly folic acid and vitamin B_{12} (cobalamin). Whether this is cause or effect is unknown, although many patients do show some improvement when they are given large doses of B vitamins. Ginkgo biloba also seems to help some patients.

Choline is a nutrient needed by the brain to produce the neurotransmitter acetylcholine. Although choline is not a vitamin—your body can synthesize what it needs from other sources—it behaves in many ways like a B vitamin. Some Alzheimer's patients have benefited from taking choline in large doses under medical supervision. Choline is a component of the dietary fat lecithin, which is found in egg yolks and other sources. This may be yet another good reason to eat eggs (see chapter 9).

Daily Supplements for AD

If you are at risk for Alzheimer's disease (see below) or are in the earliest stages of the disease, it's possible that taking supplements of brain nutrients and antioxidants could prevent its onset or slow its progress. Here's what I suggest in addition to two daily capsule of GSH 250 Master Formula:

Vitamin B complex	100 mg
Folic acid	400 to 800 mcg
Vitamin B_{12} (cobalamin)	1,500 mcg
Vitamin C	3,000 mg
Vitamin E	800 IU
Mixed carotenes	25,000 to 50,000 IU
NAC	1,500 mg
Lipoic acid	250 mg
Glutamine	500 to 1,000 mg
Selenium	200 mcg
Zinc	30 to 60 mg
Ginkgo biloba	150 mg
Phosphatidyl serine (PS)	300 to 600 mg

The mineral manganese is needed for the production of glutamine synthetase, an essential brain enzyme. Manganese deficiency sometimes mimics senility or Alzheimer's disease. Only trace amounts of manganese—just a few milligrams a day—are needed to maintain normal brain function. Too much could worsen symptoms in someone who actually does have Alzheimer's. Good food sources of manganese include spinach, whole grains, tea, raisins, and pineapple.

There is no diagnostic test for Alzheimer's disease, but scientists are coming closer to accurate ways for predicting who is at risk. AD is much more common among people who carry a variant of the APOe gene called E4, although not everyone with the gene variant will get AD, and not everyone with AD has the variant. However, recent research shows that apparently healthy people who have the E4 variant gene show significantly reduced cell activity in the parts of the brain associated with AD plaques—a strong indicator that they will develop symptoms.

Older people are often concerned that normal minor memory loss is an early sign of Alzheimer's disease. In fact, it rarely is. Abrupt and uncharacteristic mood swings are a much more common early sign. Forgetfulness in AD patients is very different from normal forgetfulness. The Alzheimer's patient forgets entire experiences—he or she may not remember eating lunch, for example, and might want to eat again soon after. AD patients don't usually remember later even when reminded and often refuse to acknowledge that they have forgotten. Their memories deteriorate, and they are increasingly unable to compensate by writing notes or using lists.

Because caring for themselves becomes more and more difficult, AD patients need a lot of help. A recent study estimated that the cost of caring for one person with AD is $47,000 a year. Since most patients live for eight to twenty years after diagnosis, Alzheimer's can be financially and emotionally very draining. For help of many kinds, contact:

Alzheimer's Association
919 North Michigan Avenue
Chicago, IL 60611-1676
(800) 272-3900

Parkinson's Disease
One out of every one hundred Americans will have Parkinson's disease (PD) by the time they reach age fifty-five; about 1.5 million Americans have it today. This slowly progressive disease is characterized by tremors or trembling of the arms and legs, stiffness and rigidity of the muscles, and slowness of movement. The symptoms occur when the brain stops producing an important neurotransmitter called dopamine.

Parkinson's disease is generally treated with a drug called levodopa (L-dopa, for short) that helps replace some of the missing dopamine. Amino acids can play a somewhat contradictory dual role in the treatment of PD. Most doctors recommend that patients taking L-dopa eat a protein-restricted diet. That's because digesting protein releases amino acids into your bloodstream. The amino acids compete with the L-dopa to cross over into the brain.

The more amino acids in the blood, the less L-dopa reaches the brain and the less effective the medication will be.

On the other hand, high doses of the essential amino acid methionine (up to 5,000 mg daily) may actually help L-dopa work better. Tyrosine is a nonessential amino acid that can be helpful as an adjunct treatment for PD, especially when it is combined with another widely used Parkinson's drug called Sinemet. Doses of 500 to 1,000 mg daily may help produce additional dopamine in the brain. Some studies suggest that the essential amino acid tryptophan may help reduce tremor in PD patients when used in conjunction with L-dopa.

Because amino acids can be both helpful and harmful to the Parkinson's patient, use them only under a doctor's supervision.

The cause of Parkinson's disease is unknown, although many researchers believe it is due to chemical imbalances in the brain. Some recent research suggests that PD could be caused by an overload of chemical neurotoxins. There is also some evidence to suggest that the imbalance could be caused by a sluggish liver that doesn't remove toxins from the bloodstream efficiently. In a classic example of how the sick can get really sicker, the accumulated toxins affect the brain chemicals, causing the symptoms of Parkinson's disease.

It's possible, then, that you could reduce your chances of developing PD by reducing your exposure to toxic chemicals. It's also possible that if you have PD, you could benefit from regularly cleansing your liver as discussed in chapter 2. Although this is by no means a cure—PD as yet has no cure—it may help stabilize your condition and keep it from worsening. Some evidence suggests that high levels of glutathione may help protect the diseased portion of the brain from further damage. If you have PD and wish to try metabolic cleansing or supplemental glutathione, talk to your doctor first.

OSTEOARTHRITIS

Arthritis is a general term meaning joint inflammation. In fact, there are more than one hundred different kinds of arthritis, but the two most common are osteoarthritis and rheumatoid arthritis.

Osteoarthritis, or painful, inflamed joints, afflicts nearly everyone over the age of fifty to some degree. Rheumatoid arthritis is more severe but also less common than osteoarthritis. It is actually a disease of the immune system and will be discussed in detail in chapter 6.

Raising your antioxidant levels can help reduce the pain and swelling of arthritis. The reason is that free radicals can be both a cause and a result of the inflammation that is causing the discomfort. The causes of arthritis still aren't fully understood, but free radicals may play a major role in beginning the process by attacking tissues in the joint. The free radicals trigger the production of the complex immune system hormones called prostaglandins. These hormones and other substances from your body's natural repair system cause swelling and pain as they rush to the damaged joint. But repairing the damage leads to the creation of further free radicals, which causes your body to send in even more prostaglandins, which in turn creates more free radicals, and so on. If you don't have plenty of antioxidants to capture the free radicals quickly when the arthritis cascade begins, you could get trapped in a cycle of pain and further damage. And even when you break the cycle, the damage to the joint is already done.

There's some strong evidence that taking antioxidant supplements can relieve arthritis discomfort as well as or even better than nonsteroidal anti-inflammatory drugs (NSAIDs) such as ibuprofen. In particular, vitamins A, C, and E and the mineral selenium have been found to be very useful. Glucosamine sulfate, chrondroiten sulfate, omega-3 fatty acids, and shark and bovine cartilage are other antioxidant supplements that give relief to some people. Glutathione has also been shown to help prevent free radical damage from prostaglandins. In my experience, the combination of antioxidant vitamins and glutathione is much more effective for arthritis than just the vitamins alone.

One big advantage of antioxidants over drugs is that they are basically safe and have very few side effects. A steady diet of NSAIDs, even in nonprescription doses, can cause gastritis and stomach ulcers. It is also a frequent cause of damage to the lining of the small intestine, which can lead to leaky gut syndrome. As we discussed in chapter 2, leaky gut syndrome can worsen your arthritis symptoms by causing autoimmune reactions that attack the affected joint. (We'll talk about this more in the discussion of autoimmune problems in chapter 6.) This is a good example of how standard drug treatment can actually make a problem worse, not better.

As a chiropractor, I have many patients who suffer from osteoarthritis. Brenda, a forty-year-old bank executive, recently came to see me after she heard me speak about glucosamine sulfate and glutathione on my Saturday morning radio show. For several years she had been taking large doses of nonsteroidal anti-inflammatory drugs to treat her arthritis. The doses were causing serious gastrointestinal discomfort, however, and she wanted to stop taking the drugs. Two weeks before she came to me, she had begun to take 1,500 mg of glucosamine daily along with three capsules a day of Glutaplex (each capsule contains 50 mg each of NAC, lipoic acid, and GSH). When we talked at my office, she told me that her symptoms had already decreased by 25 percent. We decided to add more protein to her diet and put her on a supplement schedule similar to the one below. Over a year, her symptoms improved 90 to 95 percent by her own evaluation. She continues to keep her symptoms under control by taking somewhat smaller maintenance doses of her supplements.

Daily Supplements for Arthritis

Many of my other patients have also improved, some of them to an amazing degree, when they started taking antioxidant supplements. Here's what I suggest you try if you have osteoarthritis:

Vitamin A	5,000 IU
Vitamin C	1,500 mg
Vitamin E	600 IU
Mixed carotenes	20,000 IU
GSH	250 mg
NAC	200 mg
Selenium	100 mcg
Glucosamine sulfate	1,500 mg

Omega-3 fatty acids   3 capsules
Chrondroiten sulfate   500 mg

Many of the hormones, enzymes, and other substances your body makes as part of the natural repair process are later broken down in your liver for recycling and excretion. As discussed in chapter 2, if your liver isn't functioning efficiently, the waste products won't be disposed of promptly. Instead, they and the free radicals they create may continue to circulate in your body, causing additional damage. Also, if you are producing lots of anti-inflammatory substances because you are having an arthritis flare-up, you may be making more free radicals than your liver can handle, even if it is functioning properly. Many of my patients who have tried a metabolic cleansing program report that their osteoarthritis symptoms are much better afterward.

NATURAL TREATMENTS FOR HIGH CHOLESTEROL

Researchers are actively studying some interesting new natural treatments for high cholesterol. One promising area of interest is tocotrienols. These substances are constituents of vitamin E (also known as tocopherol), but their chemical structure is slightly different. The tocotrienols, particularly one called alpha tocotrienol, seem to be even better antioxidants than vitamin E. In animal studies, alpha tocotrienol has been shown to reduce cholesterol levels. In human studies, alpha tocotrienol seems to help prevent oxidation of LDL cholesterol and to lower serum cholesterol levels. Since oxidation by free radicals is the first step in the formation of arterial plaque, it's possible that alpha tocotrienol supplements may be useful for treating high cholesterol and preventing heart disease. Alpha tocotrienol is found in palm oil, rice bran, and barley oil; 200 mg supplement capsules are available from Solgar and Allergy Research Group.

Another area that shows great promise for treating high cholesterol is plant isoflavones (also called phytoestrogens). These complex, naturally occuring compounds have powerful antioxidant effects. Soybeans are very rich in isoflavones and are the only source of one called genestein. Recent research suggests that genestein may help reduce or prevent atherosclerosis by its potent antioxidant effect and perhaps by causing you to excrete more cholesterol in your bile. Genestein may also make LDL cholesterol less sticky and therefore less likely to form deposits in your arteries; it may also make blood platelets less sticky, reducing your chances of a blood clot. Some studies suggest that adding soy isoflavones to your diet could naturally decrease your cholesterol levels by as much as 35 percent. Supplements containing mixed soy isoflavones, including genestein, are available from Solgar.

An interesting new natural remedy for high cholesterol is an herb from India called googul. According to Jerry Hickey of Hickey Chemists, a leading health food emporium in New York City, googul could help decrease your total cholesterol, decrease your LDL cholesterol, and increase your HDL cholesterol. Capsules containing 250 mg of googul are available under the brand name Googulplex from Phytopharmica. Take three capsules daily with food.

REFERENCES

Cutler, R. G. "Antioxidants and Aging," *American Journal of Clinical Nutrition*, vol. 53, no. 1 (supplement), pp. 373S–79S, January 1991.

Evans, William, and Irwin H. Rosenberg. *Biomarkers: The 10 Determinants of Aging You Can Control* (New York: Simon and Schuster, 1991).

Fahn, S. "An Open Trial of High-Dosage Antioxidants in Early Parkinson's Disease," *American Journal of Clinical Nutrition*, vol. 53, no. 1 (supplement), pp. 380S–82S, January 1991.

Ferrari, R., C. Ceconi, et al. "Role of Oxygen Free Radicals in Ischemic and Reperfused Myocardium," *American Journal of Clinical Nutrition*, vol. 53, no. 1 (supplement), pp. 215S–22S, January 1991.

Flaherty, J. T. "Myocardial Injury Mediated by Oxygen Free Radicals," *American Journal of Medicine*, vol. 91, no. 3C, pp. 79S–85S, September 30, 1991.

Fracarelli, M. et al. "Acute Effects of Carnitine in Primary Myopathies Evaluated by Quantitative Electromyography," *Drugs Experimental and Clinical Research*, vol. 6, pp. 413–20, 1984.

Georgakas, Dan. *The Methuselah Factors* (Chicago: Academy Chicago Publishers, 1995).

Hayflick, Leonard. *How and Why We Age* (New York: Ballantine Books, 1994).

Hellerstein, Herman. *Healing Your Heart* (New York: Simon and Schuster, 1990).

Julius, Mara. "Glutathione and Morbidity in a Community-Based Sample of the Elderly," *Journal of Clinical Epidemiology,* vol. 47, no. 9, pp. 1021–26, 1994.

Kidd, Parris M. "Phasphatidyl Serine and Aging," *Health & Healing,* January 1996.

Kudchodkar, B. J., et al. "Effects of Plant Sterols on Cholesterol Metabolism in Man," *Atherosclerosis,* vol. 23, pp. 239–48, 1976.

Lehr, David. "A Possible Beneficial Effect of Selenium Administration in Antiarrhythmic Therapy," *Journal of the American College of Nutrition,* vol. 13, no. 5, pp. 496–98, 1994.

Luc, G., and J.-C. Fruchart. "Oxidation of Lipoproteins and Atherosclerosis," *American Journal of Clinical Nutrition,* vol. 53, no. 1 (supplement), pp. 206S–9S, January 1991.

Marcolongo, R., et al. "Double-Blind Multicentre Study of the Activity of S-adenosyl-methionine in Hip Osteoarthritis," *Current Therapeutic Research,* vol. 37, pp. 82–94, 1985.

Null, Gary. *Reverse the Aging Process Naturally* (New York: Villard Books, 1993).

Passwater, Richard A. *The Antioxidants* (New Canaan, Conn.: Keats Publishing Inc., 1985).

Potter, J. D., et al. "Soy Saponins, Plasma Lipids, Lipoproteins and Fecal Bile Acids: A Double-Blind Crossover Study," *Nutrition Reports International,* vol. 22, pp. 521–28, 1980.

Rosenfeld, Albert. *Prolongevity II* (New York: Knopf, 1985).

Singh, Ram. "Effect of Antioxidant-Rich Foods on Plasma Ascorbic Acid, Cardiac Enzyme, and Lipid Peroxide Levels in Patients Hospitalized with Acute Myocardial Infarction," *Journal of the American Dietetic Association,* vol. 95, no. 7, pp. 775–80, July 1995.

Suarna, C., R. L. Hood, et al. "Comparative Antioxidant Activity of Tocotrienols and Other Natural Lipid-Soluble Antioxidants in a Homogenous System, and in Rat and Human Lipoproteins," *Biochem Biophys Acta,* vol. 1166, nos. 2–3, pp. 163–70, February 1993.

Suzuki, Y. J., M. Tsuchiya, et al. "Structural and Dynamic Membrane Properties of Alpha-tocopherol and Alpha-tocotrienol: Implication to the Molecular Mechanism of Their Antioxidant Potency," *Biochemistry,* vol. 32, no. 40, pp. 10692–99, October 1993.

Walford, Roy L. *Maximum Life Span* (New York: W. W. Norton, 1983).

Watkins, T., P. Lenz, et al. "Gamma-tocotrienol as a Hypocholesterolemic and Antioxidant Agent in Rats Fed Atherogenic Diets," *Lipids,* vol. 28, no. 12, pp. 1113–18, December 1993.

Weindruch, Richard. "Caloric Restriction and Aging," *Scientific American*, pp. 46–52, January 1996.

Wilcox, J. N., et al. "Thrombotic Mechanisms in Atherosclerosis: Potential Impact of Soy Protein," *Journal of Nutrition*, vol. 125 (supplement), pp. 631S–38S, 1995.

Witztum, J. N., et al. "Role of Oxidized Low-Density Lipoprotein in Atherosclerosis," *British Heart Journal*, vol. 69 (supplement), pp. S12–S18, 1993.

Yu, Byung P., ed. *Free Radicals in Aging* (Boca Raton, Fla.: CRC Press, 1993).

C H A P T E R 6

GLUTATHIONE AND YOUR IMMUNE SYSTEM

Of all the complex workings of your body, your immune system is perhaps the most amazing. It protects you against infection from harmful pathogens such as bacteria and viruses; it protects you against cancer; and it removes cellular debris from your body. To do all this, your body produces an array of specialized immune system cells, hormones, and other substances. But to manufacture all these components and keep them functioning well, your body needs the right levels of proteins, micronutrients, and antioxidants. If you can't provide your immune system with the nutrients it needs, you become much more vulnerable to illness.

Fortunately, good nutrition can be a big help in improving your immune function. If you seem to spend all winter nursing a never-ending cold, for example, it's possible that changing your diet and adding nutrients and antioxidants could boost your immune system and help cure your cold quickly and for good. In my practice I often see patients who seem to always be sick with some minor, nagging illness or infection. When I discuss their diet with them, it quickly becomes clear that what they eat—or don't eat—is contributing to their poor health. Dietary changes and supplements often return these patients to a much better level of overall well-being. In particular, they find that supplemental GSH or cysteine, along with antioxidant vitamins and trace minerals, are very helpful. One of my patients, Jeanne, is a good example. Every year, Jeanne averaged four or five colds that each lasted for weeks. She would finally recover from one cold only to get another a few weeks later. Her doctor told her to eat better, get more sleep, and exercise moderately, but he wasn't interested in providing any specifics.

When Jeanne came to me and we did a nutritional analysis of her diet, it was clear that her intake of antioxidants from food was severely lacking. She ate almost no fresh vegetables, very little fruit, and took no supplements. To improve Jeanne's antioxidant status, she added two salads and three pieces of fresh fruit to her usual daily diet. She also began taking 1,000 mg of vitamin C daily along with a high-quality multivitamin/mineral supplement and supplements of Glutathione 250 Master Formula.

Within a few days of changing her diet, Jeanne's lingering cold had cleared up. She didn't catch another one for several months; when she did, it went away in a few days. Today, Jeanne still catches a cold once or twice a year, but she doesn't get as sick and the colds clear up quickly and completely.

Good nutrition doesn't just help minor ailments such as colds. Many researchers believe that nutritional support can help treat both immune and autoimmune illnesses. Antioxidants and GSH may help boost immune function in HIV and AIDS patients, for example. In addition, keeping your immune system functioning at a high level may help keep you from getting cancer.

UNDERSTANDING YOUR IMMUNE SYSTEM

The elegant intricacy of your immune system involves your skin, mucus membranes, lymphatic organs (particularly your thymus and lymph nodes), and bone marrow.

External Barriers

Your skin and the mucus membranes of your respiratory and intestinal tracts are your body's first line of defense against illness. If your skin is in good condition, there will be no openings that allow bacteria to enter. Of course, you could cut yourself accidentally, but in general, skin that is smooth and supple resists invasion, while chapped, dry, cracked, or damaged skin has microscopic openings that provide entry for pathogens. Similarly, if the mucus linings of your lungs, stomach, and intestines are healthy, it will be hard for bacteria or viruses to enter. You can help keep your skin and mucus membranes healthy and well hydrated by drinking six to eight 8-ounce glasses of water or other liquids a day. If the air around you is very dry, I strongly recommend using a humidifier to add moisture. Cigarette smoke, air pollution, dust, and other air-borne particles can damage the linings of the lungs and allow harmful pathogens to enter. If you smoke, stop. Air filters or air conditioners may help reduce the number of particles in the air, but be sure to clean the filters often.

The Thymus

Your thymus is of crucial importance to your immune system. This walnut-size gland is found in your upper chest, just below your thyroid and above your heart. The thymus produces T lymphocytes, which are specialized white blood cells that help you fight off infections and suppress cancerous cells. The thymus also produces important hormones that help regulate many immune system activities.

At birth, your thymus is relatively large. As you get older, it gradually gets smaller. Up to a point, thymus involution (shrinkage) is perfectly normal and can't be prevented. All your glands are susceptible to free radical damage, however, and the thymus is particularly sensitive. If your redox operations are out of balance from illness, stress, environmental toxins, or poor nutrition, your thymus will be exposed to additional free radicals. The damage caused by these destructive molecules could lead to excess involution and a shortage of T cells and thymus hormones. The result? Low immunity.

The Lymphatic System

The spaces between the cells of your body are filled with a fluid called lymph. Among other roles, lymph removes wastes from your cells. Lymphatic fluid is carried through your body in lymphatic vessels that lead to lymph nodes, which are collection points for the lymph. Within the nodes, macrophages and B-cell lymphocytes remove cellular debris and pathogens. You have lymph nodes in your throat, armpits, and groin. That's why you often have "swollen glands" when you are sick with a cold or flu—your lymph nodes are filled with lymphocytes working hard to remove the virus from your system. If your lymphatic system isn't functioning well, your body won't be removing wastes and fighting off infections well.

Good nutrition and free radical protection will also help your lymphatic system. Exercise is another important way to keep the lymphatic system working well. Unlike your blood, which is pumped through your body by your heart, your lymph is pushed through your body only by the movement of your body. Regular exercise—even a daily walk of just twenty minutes or so—can help you be healthier: It filters more lymph through the nodes for waste and pathogen removal. On the other hand, people who exercise very strenuously (training for a marathon, for example) frequently have a diminished immune response, probably because of the excess free radicals they produce. If you exercise that hard, discuss your diet with a sports nutritionist. You probably need to take extra antioxidants and should be sure to get enough protein.

Bone Marrow

All the white blood cells in your body originally come from specialized cells called stem cells in your bone marrow. Some, called B cells, remain in the bone marrow and mature there. Others, the T cells, migrate to your thymus and mature there. Although these cells have a common origin, they will mature and specialize to perform interconnected but very different functions.

WHITE BLOOD CELLS

Also called leukocytes, your white blood cells are your body's main defense against invading pathogens and cancerous cells. (Your red blood cells are mostly involved in carrying oxygen to your cells and transporting carbon dioxide away—they're red because they contain hemoglobin.) All white blood cells, whether they originate in your bone marrow or your thymus, circulate throughout your body in the blood and lymph. White cells are also found in lymphatic organs such as your spleen and lymph nodes, and in the mucus linings of the intestines and lungs.

Your body manufactures three different types of leukocytes: lymphocytes, monocytes, and granulocytes. These cells work together to protect you against a wide range of attackers. They communicate with one another via complex chemical messengers.

Lymphocytes

Your lymphocytes are the central component of your immune system, although they make up only about 20 to 25 percent of your white blood cells. In addition to the T and B cells we've already discussed, your body also makes a type of lymphocyte called natural killer (NK) cells. Lymphocytes are part of your acquired immune system—that is, they "remember" the characteristics of a particular virus or bacteria and can mount a faster, more effective response the next time they encounter it.

T cells, produced in your thymus, are responsible for attacking all foreign cells in your body, a process sometimes called cell-mediated immunity. The T cells are specialized into three types in order to accomplish their task: Helper T cells (also called T-4 cells) begin the immune response for both T cells and B cells. Helper T cells behave much like a general on the battlefield. They determine the presence of the enemy and send chemical messages to the defense troops—in this case, other T cells and B cells—that tell them where and how to go into action. They also create extra troops by reproducing rapidly. The actual attack on the abnormal or invading cells is carried out by killer (cytotoxic) T cells and by the B cells. The B cells kill the invaders by creating antibodies specifically tailored to destroy them. (B cell activity is sometimes called humoral immunity.) Once the invasion has been stopped, suppressor T cells send chemical messages telling the killer cells to stop.

Helper T cells and some B cells are long-lived and can remember the characteristics of invaders. The next time a particular virus tries to attack again, for example, the helper T cells remember how they defended your body and quickly send out chemical messages to the B cells to attack with the same antibodies that worked before.

B cells produce several different types of antibodies or immunoglobulins (Ig). The most important is IgG, which is found in your blood and lymph. IgA is found in bodily secretions such as tears, breast milk, saliva, and the mucus linings of your intestines and lungs. IgE antibodies are related to allergic reactions and defending your body against parasites.

Natural killer cells are lymphocytes that don't seem to be completely under the control of the T cell system. NK cells attack and kill tumor cells, but they don't need to be activated by helper T cells.

Monocytes

Monocytes and their close relatives, macrophages, are white blood cells that have a large nucleus. They make up roughly between 5 and 8 percent of your white blood cells. Monocytes circulate in the blood, while macrophages live in your lymph nodes, liver, spleen, and lungs. Monocytes and macrophages are essential to fighting off general infection and removing debris from your system. They mostly accomplish this in a fairly straightforward way: They engulf and eat the cells or debris. Macrophages also release chemical messengers that attract or activate other white blood cell components such as helper T cells.

In addition, macrophages defend your body against mutated cells that could become tumors. To attack these cells, the macrophages release toxic chemicals such as hydrogen peroxide. This beneficial process, however, also releases free radicals that must be quickly mopped up before they cause unwanted cell damage. The same hydrogen peroxide that kills a potentially cancerous cell, for example, could also cause damage to the delicate membrane of a nearby normal cell, which in turn could cause it to mutate. Glutathione, superoxide dismutase, and other antioxidants that circulate in your blood and lymph are essential to keep your immune system's natural functions from accidentally causing harm.

Granulocytes

Most of your white blood cells are granulocytes. These cells contain tiny granules and have lobed nuclei. They fall into four main categories. Neutrophils are the largest group of white blood cells— they make up about half or more of the total. These cells attack invaders much as monocytes do, by digesting or dissolving them. If you have an infection or illness or are getting over one, the number of neutrophils in your blood is markedly higher. Eosinophils make up only about 2 to 4 percent of your white blood cells. They are particularly important for fighting off parasites and helping to stop allergic reactions. Basophils are tiny white blood cells that make up less than 1 percent of the total. They are probably responsible for releasing histamine (a vasodilator), heparin (an anticoagulant), and other substances such as vasoconstrictors into the blood, often as part of an allergic response. Basophils circulate in your blood. Mast cells, which are closely related to basophils, stay put in your tissues. They are capable of releasing large amounts of histamine during an allergic response.

Lymphokines and Monokines

The different types of immune system cells communicate with one another through complex chemical messengers; these chemicals also destroy the invading cells. Lymphocytes produce chemical messengers/attackers called lymphokines. Some of the lymphokines that have been identified so far include interferon and interleukin. The monocytes use chemicals called monokines to communicate and attack. When a macrophage engulfs a bacteria, for example, it releases chemicals that tell helper T cells what antibodies to mobilize for the attack. The macrophage's chemical messengers attract more macrophages and other white blood cells to the battle.

BOOSTING YOUR IMMUNE SYSTEM

A strong immune system protects you from illness, but only if you support and protect it. You have more than a trillion lymphocytes in your body when you're healthy, and you create millions more when you're sick. As part of normal living, you must constantly make more immune cells and dispose of the worn-out ones, at the rate of about 200,000 new immune cells every second of your life—to say nothing of the constant need to create chemical messengers and new antibodies. It's clear that good nutrition, with enough protein, vitamins, micronutrients, and antioxidants, is essential for keeping your immune system up to strength.

All the essential amino acids, which come from the proteins you eat, are the building blocks your body uses to build white blood cells and make the complex messenger proteins. You also need the amino acids, particularly cysteine, to make the antioxidant enzymes glutathione and superoxide dismutase.

Vitamins play an extremely important role in keeping your immune system functioning well. Your skin, lungs, and intestinal tract need high levels of carotenes to function well, for example. You also need carotenes because you will have fewer T cells if you have too little vitamin A, which your body manufactures from carotenes. Antibody production needs plenty of B vitamins, especially B_6 (pyridoxine) and B_{12} (cobalamin). Your white blood cells, especially the macrophages, function best when they are richly supplied with vitamin C. Vitamin E also helps improve your immune system function. Studies have shown that older people with high vitamin E levels are less likely to get infections.

The trace mineral zinc is essential to strong immune function. Zinc improves the effectiveness of your T cells and macrophages. Without enough of it, your thymus and lymph nodes shrink and your T cell count is reduced. Because your body can't really store zinc, it's very important to get an adequate daily dose of about 15 to 30 mg a day (the RDA for zinc is 15 mg). Good dietary sources of zinc include nuts, meats, fish, chicken, oysters, split peas, whole grains, and eggs. Even if you eat an adequate diet, however, your zinc levels could be too low. Taking oral contraceptives, pregnancy, obesity, chronic illness, or constant stress can all lower your zinc levels to the danger point. My patients often respond dramatically when we add zinc in the form of zinc picolinate or zinc citrate to their diet.

You also need adequate selenium levels to produce enough antibodies efficiently. Since you also need adequate selenium levels for optimum glutathione utilization, be sure you're getting about 100 mcg daily.

Although metals such as zinc and selenium are vital to your health, some other metals in your body are very undesirable. In particular, lead and copper have a weakening effect on your T cells. Excess copper in the body is quite rare, but many people have too much lead (see chapter 3 for more information).

All the amino acids are crucially important to your immune system. Your body uses these vital protein building blocks to nourish the white blood cells and to make antibodies and chemical messengers. Among the essential amino acids, lysine, methionine, phenylalanine, and tryptophan are the most important for T cells and antibody production. Human white blood cells are very high in glutathione. In particular, macrophages need plenty of glutathione. Without it, they are less effective in finding and destroying invaders.

The normal activities of your immune system create numerous free radicals; even more are created when you are fighting off an infection or illness. Paradoxically, free radicals can damage your thymus gland and keep it from creating T cells properly. This in turn could lead to reduced immune function just when you need your immune system the most.

Supplements to Support Your Immune System

You can help prevent free radical damage to your thymus by eating a nutritious diet with plenty of fresh fruits and vegetables and by taking the basic supplements discussed in chapter 1. If you're sick or have an infection, you need additional supplements to support your immune system as it works harder to heal you. In particular, you need to be sure you're getting enough vitamins, zinc, selenium, and protein. You also need extra antioxidants to protect you from your increased level of free radicals. When you have a minor illness such as a cold, try taking these supplements:

Vitamin A	10,000 IU
Vitamin B_6	50 mg
Vitamin B_{12}	2,000 mcg
Pantothenic acid	250 mg
Vitamin C	2,000 to 5,000 mg
Vitamin E	200 IU
Mixed carotenes	25,000 IU
Zinc	15 mg

| Selenium | 100 mcg |
| NAC | 400 mg |

I also suggest taking two capsules daily of GSH 250 Master Glutathione Formula from Douglas Labs until the illness has passed.

Some herbal preparations may help give your immune system a boost when you are sick. Echinacea, for example, has long been recommended by herbal healers for preventing and treating colds; it is also often suggested for respiratory and urinary tract infections. Careful scientific study indicates that echinacea probably works by stimulating the activity of macrophages. If you wish to try echinacea to treat a cold, sore throat, flu, or other minor ailment, I recommend taking an alcohol-based fluid extract from a reputable manufacturer. The usual dosage is 10 to 15 drops every few hours. Garlic may also stimulate the immune response. If you want to try this, I suggest taking two odorless garlic capsules (Kyolic is a good brand) three times a day.

Eating sugar can have a very negative effect on your white blood cells. Studies have shown that ingesting just 100 g (only about 3.5 ounces) of any type of sugar—glucose, fructose, sucrose, honey, or even fruit juice—can lower the effectiveness of your neutrophils and keep them from destroying bacteria. Sugar can also keep your lymphocytes from functioning effectively, perhaps because it blocks vitamin C from entering. A high-fat diet also has a negative effect on your immunity. People with high levels of cholesterol, triglycerides, and free fatty acids—all the result of a high-fat diet—often have decreased white cell function and a shortage of neutrophils.

Clearly, you should avoid sugar and fats when you are sick. But since sugar has a depressive effect on your immune system even when you are healthy, it might be best to avoid it even when you aren't sick. The average American takes in more than 150 g of sugar a day, an amount that often leads not only to depressed immune function but also to obesity. Many of my patients finally shake off nagging minor infections when they eliminate sugar and high-fat foods from their diet. They feel better and they lose weight as well.

Some health care practitioners recommend fasting if you have a minor illness. As we discussed above and in chapter 2, however, fasting is not a good idea. Don't force yourself to eat if you're not feeling well, but do try to drink lots of liquids. Pure water is always good, but diluted fruit juice, mild herbal teas, and the like are also fine. If you wish, mix in some powdered free-form amino acids to make sure your immune system is getting the raw materials it needs to make white blood cells.

If you've had an illness such as a bad cold, flu, or bronchitis, you may find that you feel tired and depressed, even though you don't have any obvious symptoms. At that point, you might want to try a short metabolic cleansing program as outlined in chapter 2. Many of my patients find that this helps them regain their energy.

I believe strongly that diet and supplements can do wonders to prevent and cure illness. There are times, however, when your immune system needs the help that antibiotic drugs can provide. If you suddenly become very sick, if you have been sick for several days and don't seem to be getting better or are getting worse, or if you have an infection that is very painful, swollen, has a lot of pus, or is causing red streaks in the area, see your doctor as soon as possible.

STRESS AND YOUR IMMUNE SYSTEM

One of the most exciting new areas of scientific research today is in the field of psychoneuroimmunology. Simply put, what you think and feel has a profound effect on your immunity. Although we divide the body into systems for the sake of convenience, every part of your body is closely interconnected to every other part. The incredibly complex chemistry of your brain is not separate from the equally complex chemistry of the rest of your body. You've probably noticed, for example, that you get sometimes get sick when you are under unusual stress. Why? Because stress causes the part of your brain called the hypothalamus to release a hormone that in turn stimulates your pituitary gland to release a hormone that in turn stimulates the adrenal glands on top of your kidneys to release steroid hormones into your bloodstream. Your T cells, however, are very sensitive to this particular type of steroid hormone and are damaged by it. In fact, if you are under enough stress, your thymus gland will actually shrink. In short, the cascade of hormones released by stress has depressed your immune system.

Stress—both negative and positive—is a part of life. Since you can't always avoid it, you must learn instead to deal with it. Throughout this book I emphasize the need for a good, nutritious diet, a clean environment, and moderate exercise. To those recommendations I would add getting plenty of rest (at least seven hours of sleep a night) and regular relaxation.

IMMUNE SYSTEM DISEASES

Sometimes your immune system stops being able to tell the difference between you and an invader. If this happens, your body could start to attack its own cells, causing you to develop an autoimmune disease. Autoimmune diseases such as lupus erythematosus are still poorly understood, but current research suggests that nutritional status and antioxidants may play an important role in both *causing* and *treating* them.

Lupus Erythematosus
Lupus erythematosus is a chronic autoimmune disease that causes inflammation of various parts of the body, especially the skin, joints, blood, and kidneys. Between 1.4 and 2 million Americans—1 out of every 185—have been diagnosed with lupus. For unknown reasons, about 80 percent of all people with lupus are women. If you smoke, you are more likely to get lupus.

One type of lupus, called discoid lupus, is limited only to the skin. Patients get a red rash on the face, neck, and scalp that can leave permanent scars. Systemic lupus erythematosus (SLE) is usually more severe than discoid lupus and can affect almost any organ or system of the body. The most common symptoms include achy or swollen joints, fever, prolonged fatigue, skin rashes, anemia, kidney problems, pleurisy, and sensitivity to sunlight.

The causes of lupus are unknown, although two areas of research show some promise. Many researchers believe that a virus triggers lupus. The patient's immune system produces antibodies to fight off the virus, but for some reason, it continues to produce the antibodies even after the virus has been eliminated; the antibodies then attack the patient herself. Support for this theory comes from the fact that treatments such as low-level radiation and some anticancer drugs, which suppress the immune system, often benefit lupus patients.

Another possible cause of lupus is yeast overgrowth of the small intestine and bowel (see chapter 2). Some researchers think that toxins from the yeast overgrowth could enter the bloodstream and cause decreased suppressor T cell function. If the immune system can't produce enough suppressor cells, it can't tell the B cells to stop producing antibodies once the viral threat has been eliminated.

People with lupus often have periods of remission alternating with flare-ups. Unfortunately, what causes this up-and-down aspect of the disease is still largely unknown. There is some evidence that diet can play a role, although research is still in the early stages. Many lupus patients have reported flare-ups of symptoms after eating particular foods, particularly plant foods including soy beans, corn, spinach, and carrots. It's possible that these foods contain a plant protein that could trigger autoimmune antibodies in lupus patients.

In one study, a volunteer who ate alfalfa as part of a cholesterol-lowering study developed lupus symptoms. The effect was later duplicated in studies with monkeys. The amino acid L-canavanine was identified as the culprit. Lupus patients should avoid alfalfa sprouts and any health food products that contain alfalfa.

The omega-3 fatty acids found in fish oils seem to show promise as a way to help control lupus symptoms. Studies have shown that high doses of omega-3 in capsule form can help reduce joint inflammation—although no more than aspirin does, and at a much higher cost. On the other hand, aspirin can sometimes trigger a flare-up.

Many lupus patients report that their symptoms of sun sensitivity and inflamed skin are reduced when they take vitamin E in doses of 800 to 2,000 IU daily. Because lupus causes inflammation, which in turn causes the release of a lot of free radicals, antioxidant vitamins may be helpful. Supplements of NAC, lipoic acid, and selenium may also help boost antioxidant activity for lupus patients.

Although lupus is rarely life-threatening, it is a chronic and serious disease that needs careful medical management. Discuss any dietary changes or supplements with your doctor before you try them.

For more information about lupus, contact:

Lupus Foundation of America
Four Research Place, Suite 180
Rockville, MD 20850
(800) 558-0121

Rheumatoid Arthritis

Rheumatoid arthritis (RA) is a chronic autoimmune disease that occurs when your own antibodies attack your joints, causing pain, stiffness, and swelling. In severe cases, the joints become deformed and permanently lose their function. People with RA may also have general symptoms such as fatigue, weakness, and loss of appetite. Rheumatoid arthritis affects more than two million Americans, two-thirds of them women.

As with other autoimmune illnesses, the causes of rheumatoid arthritis are not fully understood. Free radicals may be one significant cause; food allergies, leaky gut syndrome, and yeast overgrowth of the small intestine and bowel may be other closely related causes.

Free radicals may damage the sacs in your joints that produce synovial fluid, the lubricant that helps your joints move smoothly. Without sufficient synovial fluid, the cartilage and surrounding tissue of the joint are damaged. Taking antioxidant vitamins, NAC, lipoic acid, and selenium could help capture the damaging free radicals and prevent further damage. In fact, studies have shown that RA patients tend to have low selenium levels, which in turn means that they probably have low levels of glutathione and other important antioxidants.

Some patients feel that foods from the nightshade family, including tomatoes, potatoes, eggplant, and bell peppers, make their arthritis worse. This could be because the nightshade foods contain a toxic substance called sotanine. People with rheumatoid arthritis seem to be particularly sensitive to sotanine. There is some evidence that consuming milk and dairy products triggers RA flare-ups in some people. The autoimmune response that specific food allergies may trigger could worsen or even cause rheumatoid arthritis. If you think food allergies could be causing your arthritis, work with your health care provider to identify and eliminate the problem foods.

As discussed in chapter 2, leaky gut syndrome and yeast overgrowth of the small intestine could be causes of rheumatoid arthritis symptoms. If you have a leaky gut, particles of undigested food enter your bloodstream through your small intestine, causing an immune response. Toxins from a yeast overgrowth can also trigger your immune system and cause symptoms. Ironically, the same powerful anti-inflammatory drugs that are often prescribed to treat arthritis can also cause leaky gut syndrome and actually make the problem worse. If you think a leaky gut or yeast overgrowth is causing or worsening your arthritis symptoms, reread chapter 2 and discuss treatments with your health care provider.

Many people with rheumatoid arthritis find that fasting, going on a liquid diet, or going on a vegan diet (no animal products at all) relieves their symptoms temporarily. This could well be because they suffer from leaky gut syndrome. By fasting or eating a very restricted diet, they avoid introducing undigested food particles into their system and triggering an autoimmune response. If fasting or a vegan diet helps your RA, consider the possibility of leaky gut syndrome.

Daily Supplements for RA

Patients with rheumatoid arthritis often benefit from fairly large doses of supplemental antioxidant vitamins and the trace minerals zinc and selenium, along with supplemental NAC and lipoic acid. It's also important for them to get enough calcium to prevent bone loss. Here's what I recommend on a daily basis:

Vitamin C	2,000 mg
Vitamin E	400 IU

Mixed carotenes	25,000 IU
Pantothenic acid	250 mg
Niacinamide	500 to 1,000 mg
Calcium	1,500 mg
Selenium	200 mcg
Zinc	50 mg
Magnesium	1,000 mg
NAC	600 to 1,200 mg
Omega-3 fatty acids	1 to 3 capsules
Shark cartilage	3 capsules
Sea cucumber	3 capsules
Superoxide dismutase	3 capsules
Glucosamine sulfate	1,500 mg

If you prefer a more convenient multiple supplement, try Arthrin from Applied Nutrition Technology. No matter how you take your supplements, I also suggest a daily drink made from free-form amino acid powder mixed with pure water to provide extra protein for tissue repair.

Some of my RA patients find that taking omega-3 fatty acids (fish oil) or gamma linoleic acid (GLA), a component of unsaturated fats found in plants such as black currant, borage, and evening primrose, helps relieve their symptoms. Fish oil and GLA affect the body's production of prostaglandins, hormone-like substances that play a role in inflammation. If you'd like to try either of these, start with small daily doses of just one capsule and gradually increase the dose to no more than three (cut back if you get heartburn). Check with your doctor before using fish oil if you have diabetes or any sort of blood clotting problem.

Multiple Sclerosis

Multiple sclerosis (MS) is a chronic disease of the central nervous system that leaves its victims with varying degrees of interference with speech, walking, and other basic functions. The symptoms of MS are caused when the fatty sheath that forms a covering for the nerve fibers of the central nervous system is destroyed. The impulses that normally travel along the nerve are disrupted, much as removing the insulation around an electrical wire interferes with the transmission of signals. About 350,000 Americans have MS, with nearly two hundred new cases diagnosed every week. Multiple sclerosis often strikes people in their prime, most commonly between the ages of twenty and forty. Nearly twice as many women as men have MS.

The cause of MS is still unknown. Many researchers believe that a virus is the primary cause, others feel that it is an autoimmune disease, and some feel that diet is the cause. Some MS patients do experience a viral infection or a period of poor nutrition shortly before their symptoms begin, but many others do not.

The symptoms of MS vary greatly, depending upon where the nerve damage occurs. Typically, someone with multiple sclerosis has periods of active disease, called exacerbations, and symptom-free periods, called remissions. One common feature of MS is that many symptoms worsen when patients are exposed to heat.

There is no cure for multiple sclerosis; physicians can only treat the symptoms. Numerous special diets that claim to miraculously cure or help multiple sclerosis have been tried, and although some seem to help a few people, at least for a time, none have been found to have genuine therapeutic value. These diets may seem to work largely because MS is an episodic disease. Remissions that seem to have been caused by a diet simply occurred naturally. The Swank diet, named for the physician who first proposed it in 1948, has not been accepted by the National MS Society's Medical Advisory Board as being effective, although some patients claim it has been very helpful for them. There is no firm evidence that this diet, which is very low in saturated fats and high in omega-3 fatty acids, can help treat or cure MS. However, there is some evidence to suggest that saturated fats may trigger or exacerbate autoimmune responses. If MS is an autoimmune disease, avoiding saturated fats as recommended by the Swank diet may indeed help.

Most doctors recommend that MS patients take a good daily multivitamin. It is possible that taking larger doses of antioxidant vitamins and supplemental GSH or NAC, lipoic acid, and selenium may help prevent additional nerve damage from free radicals. Before adding these supplements to your diet, however, discuss them with your doctor.

For more information about multiple sclerosis, contact:

National Multiple Sclerosis Society
Education Department
733 Third Avenue
New York, NY 10017
(800) FIGHT-MS

AIDS

Acquired immune deficiency syndrome, better known as AIDS, differs from other autoimmune diseases in that we know exactly what causes it: the human immunodeficiency virus (HIV). The virus invades the victim's T cells, especially the helper (T-4) cells, and causes a collapse of the immune system. Without a functioning immune system, AIDs patients can't fight off infections or cancer and eventually die.

Sadly, there is no cure for AIDS. As scientists learn more and more about how the virus does its damage, however, new ways to treat AIDS are being proposed. Much of the research today centers on understanding exactly how HIV depletes the T cells of the immune system. The apoptosis theory is particularly intriguing.

Apoptosis, or the destruction of a cell by breaking it up into fragments, is a normal part of your immune system. It's how your body gets rid of old or damaged cells or cells that are no longer needed. Your macrophages remove the cell fragments. Researchers now believe that HIV somehow makes the apoptosis process go haywire and causes the T cells to destroy themselves. Without helper T cells to organize the body's response to pathogens, the immune system is defenseless.

Because a low glutathione level makes cells more vulnerable to apoptosis, and because AIDS patients are often generally deficient in glutathione, many patients now take supplemental cysteine in the form of NAC. The idea is to protect the T cells against apoptosis by trying to boost the glutathione level. In addition, NAC seems to lower the blood levels of a substance called tumor necrosis factor (TNF); this substance seems to help HIV multiply, so reducing its presence in your bloodstream could slow the progression of AIDS.

For reasons that are not clearly understood, many HIV and AIDS patients have digestive absorption problems, so it is usually best for them to take NAC instead of GSH. Physicians working with AIDS patients generally recommend a fairly high dosage of NAC—1,800 to 2,400 mg daily, taken as three or four 600 mg capsules spread evenly throughout the day. To improve absorption, the capsules should be taken about half an hour before eating. The effectiveness of the NAC will generally be enhanced by taking 250 mcg of supplemental selenium, also spread over the day. Additional glutamine, vitamins E, C, and B_6, and zinc may also help raise the GSH level.

Some researchers believe that lipoic acid will prove to be even more effective than NAC in preventing apoptosis. Studies on this have only just begun, however, and it is too soon to know if it will be helpful.

If you have HIV or AIDS, nutritional approaches and natural medicine can often be very beneficial, but they must be used with caution. Too much vitamin C, for example, can cause diarrhea, which can be very dangerous to an AIDS patient. Discuss nutrition, supplements, and natural remedies with your health care provider.

REFERENCES

Baird, I., R. Hughes, H. Wilson, et al. "The Effects of Ascorbic Acid and Flavonoids on the Occurrence of Symptoms Normally Associated with the Common Cold," *American Journal of Clinical Nutrition*, vol. 32, pp. 1686–90, 1979.

Beisel, W. R. "Single Nutrients and Immunity," *American Journal of Clinical Nutrition*, vol. 35, pp. 417–68, 1982.

Beisel, W., R. Edelman, K. Nauss, et al. "Single-Nutrient Effects on Immunologic Functions," *Journal of the American Medical Association*, vol. 245, pp. 53–58, 1981.

Berger, Stuart M., M.D. *Dr. Berger's Immune Power Diet* (New York: New American Library, 1985).

Blumberg, Jeffrey B. "The Role of Vitamin E in Immunity During Aging," M. Mino et al., eds., *Vitamin E: Its Usefulness in Health and Diseases* (Farmington, Conn.: S. Karger, 1993).

Bounous, G., and P. A. L. Kongshavn. "Differential Effect of Dietary Protein Type on the B cell and T cell Immune Responses in Mice," *Journal of Nutrition*, vol. 115, pp. 1403–8, 1985.

Boyne, R., and J. Arthur. "The Response of Selenium-Deficient Mice to Candida Albicans Infection," *Journal of Nutrition*, vol. 116, pp. 816–22, 1986.

Broder, Samuel. "NAC," *The Body Positive*, vol. 2, no. 8, p. 18, November 1989.

Buhl, R., et al. "Systemic Glutathione Deficiency in Symptom-Free HIV-seropositive Individuals," *The Lancet*, pp. 1294–98, December 2, 1989.

Carroll, David L., and Jon Dudley Gorman. *Living Well with MS* (New York: HarperPerennial, 1993).

Cohen, L. "Fish Oils in Rheumatoid Arthritis," *The Lancet*, pp. 720–21, September 26, 1987.

Franulovich, Tim. "Glutathione Levels Raised with NAC," *The Body Positive*, vol. 4, no. 5, p. 14, May 1991.

Fraser, R., S. Pavlovic, C. Kurahara, et al. "The Effect of Variations in Vitamin C Intake on the Cellular Immune Response of Guinea Pigs," *American Journal of Clinical Nutrition*, vol. 33, pp. 839–47, 1980.

Hall, N., and A. Goldstein. "Thinking Well: The Chemical Links Between Emotions and Health," *The Sciences*, pp. 34–40, March 1986.

Hansen, T., A. Lerche, V. Kassis, et al. "Treatment of Rheumatoid Arthritis with Prostaglandin E1, Precursors Cis-linolenic Acid and Gamma Linolenic Acid," *Scandinavian Journal of Rheumatology*, vol. 12, pp. 85–88, 1983.

Horrobin, D., M. Mauku, and M. Oka. "The Nutritional Regulation of T-lymphocyte Function," *Medical Hypotheses*, vol. 5, pp. 969–85, 1979.

Kidd, Parris M., and Wolfgang Huber. *Living with the AIDS Virus* (Berkeley, Ca.: HK Biomedical, 1995).

Kremer, J. M., W. Jubiz, A. Michalek, et al. "Fish-oil Fatty Acid Supplementation in Active Rheumatoid Arthritis," *Annals of Internal Medicine*, vol. 106, pp. 497–503, April 1987.

Laurence, J. "The Immune System in AIDS," *Scientific American*, vol. 253, pp. 84–93, December 1985.

Link, Derek. "NIC, NAC, Paddy Whack . . . ," *Notes from the Underground*, p. 1, July/August 1992.

McCord, J. "Free Radicals and Inflammation: Protection of Synovial Fluid by Superoxide Dismutase," *Science*, vol. 185, pp. 529–31, August 1974.

Meydani, M. "Vitamin/Mineral Supplementation, the Aging Immune Response, and Risk of Infection," *Nutrition Reviews,* vol. 51, no. 4, pp. 106–9, 1993.

Panush, R., R. Stroud, and E. Webster. "Food-Induced (Allergic) Arthritis: Inflammatory Arthritis Exacerbated by Milk," *Arthritis and Rheumatology,* vol. 29, no. 2, pp. 220–26, February 1986.

Passwater, Richard A. "Lipoic Acid Against AIDS," *Whole Foods,* pp. 50–60, December 1995.

Phillips, Robert H. *Coping with Lupus* (Garden City Park, N.Y.: Avery Publishing, 1991).

Roederer, M., F. J. Staal, et al. "N-acetylcysteine: Potential for AIDS Therapy," *Pharmacology,* vol. 46, pp. 121–29, 1993.

Rouzer, C. A., W. A. Scott, et al. "Depletion of Glutathione Selectively Inhibits Synthesis of Leukotriene C by Macrophages," *Proceedings of the National Academy of Sciences,* vol. 78, no. 4, pp. 2532–36, 1981.

Sanchez, A., J. Reeser, H. Lau, et al. "Role of Sugars in Human Neutrophilic Phagocytosis," *American Journal of Clinical Nutrition,* vol. 26, pp. 1180–84, 1973.

Schmidt, K. "Antioxidant Vitamins and Beta-Carotene: Effects on Immunocompetence," *American Journal of Clinical Nutrition,* vol. 53 (supplement), no. 1, pp. 383S–85S, 1991.

Siano, Nick, with Suzanne Lipsett. *No Time to Wait* (New York: Bantam Books, 1995).

Skoldtam, L. "Fasting and Vegan Diet in Rheumatoid Arthritis," *Scandinavian Journal of Rheumatology,* vol. 15, pp. 219–223, 1986.

Weil, Andrew. *Spontaneous Healing* (New York: Knopf, 1995).

C H A P T E R   7

GLUTATHIONE AND YOUR EYES

If you were asked to name a surgical operation performed more than 1.5 *million* times a year in the United States, what would you say? You'd probably be surprised to learn that the answer is cataract removal surgery. Cataracts are just one of the vision-robbing eye conditions that glutathione may help prevent. Some very exciting research now suggests that glutathione can help prevent or treat not only cataracts but also macular degeneration, glaucoma, diabetic eye disease, uveitis, and herpes simplex eye infections. Some 13 million older Americans suffer from the gradual and incurable vision loss of macular degeneration, and millions more suffer from other preventable eye conditions.

CATARACT PREVENTION

The lens of your eye is made up of a clear protein. As you grow older, however, the proteins in the lens may start to clump together and cause a cloudy area that blocks your vision—a cataract. You are most likely to develop a cataract if you are over age seventy, although many people get them earlier.

Repeated studies have shown that almost all age-related cataracts are caused by the cumulative effect of excessive exposure to ultraviolet (UV) light. These are the invisible, very short wavelength rays in sunshine that also make your skin tan. UV rays are very active, bouncing around inside the lens of your eye, disrupting the molecules, and creating large numbers of destructive free radicals. In addition, your eyes produce free radicals as a normal part of their metabolic processes.

Your eyes are naturally well equipped to fight back and quench the free radicals as soon as they are formed. Your eye tissues and the fluids that nourish the lens and make up the inside of your eye contain very high levels of antioxidants. In fact, if your eyes are healthy, their level of vitamin C is up to twenty times as high as the level in your blood. Levels of the antioxidant enzymes GSH, catalase, and superoxide dismutase (SOD) in the eye are also high.

Years of constant bombardment with UV light takes a toll on your eyes. More free radicals are created than your natural antioxidant defenses can handle. The situation worsens as you age and the production of antioxidants slows down. As you get older, the level of glutathione in your lens drops sharply, often to a dangerously low point. This lets the free radicals gradually get the upper hand in your lens. The result? Cataract. And the lower your GSH level, the more severe the cataract will be.

Cigarette smoking and alcohol consumption can also increase your risk of cataracts. By some estimates, 20 percent of all cataracts can be attributed to smoking, and people who have more than two drinks a day also have a high chance of cataracts.

Your ophthalmologist can detect the beginnings of a cataract long before you have any symptoms. It could be months or even years later before you notice any vision changes. As the cataract gradually worsens, though, less light enters your eye. You start to need much brighter light to see properly, your night vision and peripheral vision are impaired, you can't see colors as distinctly, and you become much more sensitive to glare. Because the onset of the symptoms is usually gradual, you may not even realize you have a cataract until reading, sewing, or similar ordinary activities become very difficult. Your gradually worsening vision could also mean that you are driving dangerously, confusing your medications, experiencing falls and household accidents, or having other serious problems with everyday life.

By the time your cataract has reached the problem point, the only way to treat it is to surgically remove the clouded lens and replace it with an artificial lens. Cataract surgery today is a very

safe, painless, and effective operation that is generally performed
on an outpatient basis without general anesthesia. It is almost always covered by medical insurance. Modern microsurgery techniques mean that most patients can completely resume their
normal activities within a week or two.

No surgery is completely without risk, of course, and there's
also that long and possibly dangerous period of poor vision before
you have the operation. Wouldn't it be better to just avoid getting
a cataract to begin with?

You can. You have several practical lines of defense against cataracts. If you smoke, stop. If you have more than two drinks a day,
cut back. When you're outside in daylight, protect your eyes from
UV radiation by wearing 100 percent UV-blocking sunglasses. (See
Figure 7.1 for more information about selecting sunglasses.) You
can get additional UV protection by wearing a hat with a brim.

Most of all, keep the antioxidant levels in your eyes high. Recent studies have conclusively shown that people who have a high
dietary intake of vitamins C and E and mixed carotenes (the precursors of vitamin A) have significantly lower rates of cataract.
Studies have also confirmed that low levels of GSH in eye tissues
are directly linked to cataracts. To be sure your antioxidant levels
are high enough to maintain healthy eyes, eat a diet rich in fresh
fruits and vegetables and supplement your diet with additional
vitamins, the amino acid cysteine in the form of NAC, and other
antioxidants. The vitamins work synergistically with NAC to raise
your glutathione level. You also need all the B vitamins and the
minerals magnesium, zinc, and selenium. The plant-based substances known as flavonoids are valuable antioxidants that can also
help promote healthy eyes.

Daily Supplements for Healthy Eyes
Here's what I recommend on a daily basis for overall health and
healthy eyes:

Vitamin A	10,000 IU
Vitamin B$_1$ (thiamine)	50 mg
Vitamin B$_2$ (riboflavin)	50 mg

Vitamin B$_3$ (niacin)	50 mg
Vitamin B$_6$ (pyridoxine)	50 mg
Folic acid	400 mcg
Vitamin B$_{12}$ (cobalamin)	1,000 IU
Vitamin C	500 mg
Vitamin E	400 IU
Mixed carotenes	25,000 IU
GSH	250 mg
NAC	100 mg
Taurine	1,000 mg
Lutein	6 mg
Lemon bioflavonoids	200 mg
Rutin	100 mg
Bilberry	120 mg
Zeaxanthin	240 mcg
Magnesium	400 mg
Selenium	250 mcg
Zinc	15 mg

Many people prefer to take a multi-ingredient formula rather than a number of different tablets and capsules every day. Ocugard from Twinlife is a good choice, as is Ocucare from Nature's Plus.

If your ophthalmologist detects early cataract symptoms, you may be able to slow their progress, perhaps to the point where surgery is never needed. Follow the recommendations above regarding smoking, drinking, and UV protection. In addition to a healthy diet rich in antioxidant vitamins, I recommend the supplements discussed above, but increase the dose of vitamin E to 800 IU and vitamin C to 1,000 mg.

If you have cataracts as a result of diabetes, antioxidant supplements will almost certainly be helpful. However, it is far more important to control your blood sugar levels. If you have diabetes-related cataracts, always discuss antioxidant and amino acid supplements with your doctor before you try them. If you have cataracts from taking steroids, some types of tranquilizers, or from other drugs, discuss taking GSH and other antioxidant supplements with your doctor as well.

Many people learn they have cataracts only when their vision

Figure 7.1

GUIDELINES FOR SELECTING SUNGLASSES

- Select sunglasses that block 100 percent of ultraviolet (UV) radiation.
- UV protection comes from the lens material or coating, not the color. Darker lenses do not necessarily provide greater UV protection.
- Read labels carefully. Look for labels that state the percentage of UV protection provided. Labels that say "maximum UV protection" may not provide 100 percent protection.
- Select polarized lenses to reduce glare. Polarization does *not* provide UV protection!
- Not all mirrored sunglasses provide UV protection. Read the label.
- For best protection against UV exposure, select a wraparound design or one with side shields.
- Gray lenses provide the best natural color perception.
- Orange-brown lenses are best for people with macular degeneration.
- Yellow lenses help reduce night glare for people with cataracts.
- Avoid blue, red, or pink lenses. These choices distort your color perception.
- For added protection, wear a hat with a brim along with your sunglasses.

Source: American Optometric Association.

is affected. At that point, vitamins and amino acid supplements may still slow down cataract formation, but they won't restore your sight. Surgery is the only alternative to eventual blindness. If cataracts—even small ones—are affecting your quality of life, don't put off the surgery.

Surgery doesn't mean giving up on antioxidants, however. Antioxidant supplements can help you recover from the surgery more quickly and easily. According to Bill Sardi, an expert in nutritional approaches to eye problems, taking supplements daily for a week before the surgery and a month after it enhances wound healing and reduces inflammation.

Supplements for Eye Surgery

If you're planning to have or are recovering from eye surgery, I suggest taking these supplements:

Vitamin A	10,000 IU
Vitamin B$_2$ (riboflavin)	100 mg
Vitamin B$_6$ (pyridoxine)	100 mg
Vitamin C	500 to 1,000 mg
Vitamin E	400 IU
Mixed carotenes	10,000 to 20,000 IU
NAC	250 mg
GSH	250 mg
Glutamine	1,500 mg
Zinc	15 to 25 mg

Continue to take your supplements even after you've healed from cataract surgery. You could help prevent a cataract in the other eye. And now that your lens is clear again, you need to protect the retina at the back of your eye from the increased UV light that is again reaching it.

PREVENTING MACULAR DEGENERATION

Age-related macular degeneration (AMD) is the leading cause of vision loss in people over age sixty-five. It is caused when the macula—the central, most sensitive region of the retina—loses the yellow pigment that allows you to detect light and color. Macular degeneration destroys the sharp vision needed for seeing objects clearly. As a result, you gradually lose your ability to read, drive, recognize faces, and perform many other daily activities. About 10 percent of all people with macular degeneration lose their sight completely, while the rest experience some degree of central vision loss. About 90 percent of people with AMD have the "dry" form, which causes gradual central vision loss. The rest have "wet" AMD, a form of the disease that progresses quickly and is much more likely to lead to complete vision loss. For unknown reasons, people with wet AMD develop a rapidly growing network of excess blood vessels under the retina. The blood vessels are very fragile and break easily, which causes severe damage to the macula. Gen-

Figure 7.2

SYMPTOMS OF AGE-RELATED MACULAR DEGENERATION

- Blurred vision
- Portions of pages of a book are blurry
- Straight lines such as utility poles appear crooked, wavy, broken, or distorted
- The center of your field of vision has a small dark area or blind spot
- Some colors are distorted

Source: Prevent Blindness America.

erally, macular degeneration affects one eye first; the other eye often develops symptoms several years later. For a list of macular degeneration symptoms, see Figure 7.2.

The greatest risk factor for macular degeneration is simply age. AMD is rare among people under the age of sixty. Among people over the age of seventy-five, though, the risk of macular degeneration can be as high as 30 percent. Additional risk factors include cigarette smoking, high blood pressure, diabetes, and cardiovascular disease. If you have blue, green, or hazel eyes, you are more likely to get AMD than if you have brown or black eyes. Probably for that reason, African Americans are much less likely to develop AMD than caucasian Americans. For unknown reasons, women are more likely to get AMD than men. If you have a close relative with AMD, you are somewhat more at risk.

There's one other significant risk factor for age-related macular degeneration: nutrition. A major study in 1993 showed that older people who eat a diet high in dark green leafy vegetables such as spinach and collard greens have a 43 percent lower risk of AMD. Why? Although your macula is only about one square millimeter in size, it contains millions of light-sensing cells. When the macula degenerates, those cells lose the carotenoids lutein and zeaxanthin—the same carotenoids found in dark green leafy vegetables. Researchers think that these carotenoids help protect the macula from the harmful effects of ultraviolet light and blue-violet light from sunlight.

But why are the carotenoids lost to begin with? No one knows for sure, but free radicals caused by the ultraviolet radiation in sunlight are the most likely culprits. When sunlight hits your retina, the ultraviolet rays and light from the blue-violet end of the spectrum cause free radicals to form. Ordinarily, your eye can handle these renegades by using its high levels of vitamin C and glutathione to quench their voracious free electrons. Your eye also contains antioxidant substances called anthocyanosides, which help prevent damage to the retina from sunlight by protecting the fat molecules in the retinal cell membranes.

As you age, however, you generally produce less of these vital antioxidants. At the same time, the tiny blood vessels that supply amino acids, vitamins, minerals, oxygen, and other vital nutrients to the retina can become clogged. When the supply of antioxidant building blocks is reduced, your eye can't produce enough antioxidants to defend itself against free radicals. The result can be damage to the macula and gradual loss of vision.

There is no cure for AMD, but you can take effective steps to prevent it or slow its progression. Protect your eyes against free radicals from ultraviolet light by wearing 100 percent UV-blocking sunglasses when outside in daylight. If your sunglasses block all UV light, they also automatically block about 85 percent of blue-violet light. (See Figure 7.1 for information about selecting sunglasses.) Wear a hat with a brim for extra protection. If you smoke, stop. If you have diabetes, high blood pressure, or any other chronic condition, do your best to keep it under control.

The most important thing you can do to keep your eyes healthy and prevent AMD is to keep your levels of protective antioxidants high throughout your life. Many people who later develop AMD show early signs of it when they are still in their thirties. When they reach their fifties, their symptoms accelerate. Always be sure to eat a nutritious diet that includes plenty of leafy greens. For optimal eye health as you age, however, you need more antioxidants than you can get through diet alone—especially if you are outdoors a lot or have a close family member with AMD.

In addition to the supplements for healthy eyes recommended above, you need to boost your levels of the sulfur-containing amino acid taurine. Of all the amino acids found in your retina, taurine is the most abundant. It's essential for maintaining retinal health. Taurine is found naturally in eggs, fish, and all animal protein, particularly organ meats. You don't have to add calories, fat, or animal products to your diet to raise your taurine levels. Some people get gastric upsets from taurine, especially in doses over 100 mg. Your body readily synthesizes its own taurine from cysteine, but only if the cofactor vitamin B_6 (pyridoxine) is present as well. To make sure your eyes have enough taurine without taking supplements, you need to make sure you're getting enough cysteine and vitamin B_6. I suggest adding another 250 mg of NAC and 50 mg more of vitamin B_6.

If you are already suffering from age-related macular degeneration, you need even higher levels of antioxidants, amino acids, and cofactors—far more than diet alone could reasonably provide. The glutathione level of a person with AMD is only about half that of a person the same age with healthy eyes. Cysteine in the form of NAC is essential for boosting your glutathione level. You also need additional taurine, since the sulfur in taurine is vital for maintaining the integrity of the cell walls in the retina.

Some patients have reported that taking an extract made from bilberries, high in natural anthocyanosides, has helped their vision. Other patients claim that quercetin or ginkgo biloba extracts are helpful. The evidence for the value of these extracts is uncertain, however. You should discuss taking any supplements, especially anthocyanoside extracts, with your ophthalmologist before you try them.

There are some rare instances where supplements have apparently stopped or even reversed macular degeneration. Even so, I can't promise that taking the supplements suggested below will restore your eyes. AMD has no miracle cure. There is an excellent chance, however, that they could slow the progress of AMD and help preserve your remaining vision. Because retinal cells grow very slowly, it could take as long as a year before you notice any improvement. Stick with your program for at least that long.

To help prevent or slow macular degeneration, I suggest you take the supplements discussed at the beginning of this chapter on page 141. In addition, I recommend taking 100 mg of taurine, 250 to 400 mg of chromium, and 3,000 to 10,000 mg of anthocyanoside extract.

Even if you have central vision loss from macular degeneration, you probably still have good peripheral vision. With the help of low-vision devices, you can continue to lead an independent and active life. For more information, contact:

Prevent Blindness America (National Society to Prevent Blindness)
500 East Remington Road
Schaumburg, IL 60173
(800) 331-2020

GLUTATHIONE AND GLAUCOMA

Glaucoma is a progressive eye disease that slowly destroys the optic nerve, causing eventual loss of vision. It occurs when, for unknown reasons, the passages that drain the fluids inside your eyeball become blocked, causing a rise in the pressure inside your eye. This in turn puts pressure on the optic nerve that transmits images from your retina to your brain. The result is a painless, gradual loss of peripheral vision; in untreated cases, blindness can occur. See Figure 7.3 for a list of glaucoma symptoms.

Fortunately, more than 90 percent of all glaucoma cases are the simple open-angle type, also called common glaucoma. This type of glaucoma can be readily detected in its early stages with a routine eye exam. Glaucoma can be treated very effectively and easily, usually with medicated eye drops, but there is no cure. The treatment only keeps the condition from getting worse and must be continued for the rest of your life. More than one million Americans now have restricted vision as a result of glaucoma; about seventy thousand are blind from it.

You are most likely to get glaucoma if you are over age forty. As with many other eye problems, the older you get, the more likely it is to occur. If you have a close family member who has developed glaucoma, you are more likely to get it yourself. If you

Figure 7.3

SYMPTOMS OF GLAUCOMA

- Early glaucoma has no symptoms. Have regular eye exams.
- Elevated eye pressure can later lead to glaucoma.
- Visual symptoms usually occur only in advanced cases:
  - loss of side (peripheral) vision
  - halos or rainbows around lights
  - blurred vision
  - night blindness

Source: Foundation for Glaucoma Research.

are African American, you are at least four times more likely to develop glaucoma than if you are caucasian. In fact, African Americans between the ages of forty-five and sixty-five are fifteen times more likely to become blind from glaucoma than caucasian Americans. You're also at risk if you are very nearsighted or have diabetes.

Lifestyle choices can affect your chances of getting glaucoma. If you smoke cigarettes, if you are overweight, if you get little exercise, if you have untreated high blood pressure, or if your overall nutrition is poor, you are more likely to develop glaucoma.

Clearly, adopting a healthier lifestyle can help reduce your chances of getting glaucoma. Keeping your blood pressure and diabetes under control can help too. And though you can't change your genetic heritage or your age, you can take some nutritional steps that could help keep the odds on your side. If you have high eye pressure, adding supplemental vitamins, minerals, and amino acids to your diet along with your medication (if any) could help keep glaucoma from developing. And if you already have glaucoma, taking supplements along with your medication could help slow the progress of the disease.

Doctors still don't know exactly why the drainage passages in your eye clog when you have glaucoma. Many researchers suspect that poor blood circulation to the eye is one of the primary causes. When the tiny blood vessels in your eyes are clogged, they can't carry in the nutrients your eyes need, including vitamins and antioxidant building blocks. Among other things, this means that there are fewer antioxidants in your eyes to soak up free radicals.

Nutritional therapy for glaucoma focuses on improving blood circulation to the eyes and maintaining healthy cells within the eyes. I recommend taking vitamins E and C along with the amino acid cysteine in the form of NAC. This will help boost your ocular levels of antioxidant glutathione. I also recommend taking vitamin B_{12} supplements to help strengthen the optic nerve. If your optic nerve is healthy, it can better withstand increased eye pressure. The mineral cofactors magnesium, selenium, and zinc are also necessary.

If you have no early symptoms but are at risk for glaucoma based on your race, age, or family history, I suggest you take the basic daily supplements for maintaining healthy eyes discussed on page 141.

Promising evidence shows that a concentrated extract of the anthocyanosides found in dried blueberries or bilberries may protect against glaucoma. If you are at risk, consider adding 3,000 mg of anthocyanoside extract to your daily supplements.

There's also some evidence that omega-3 fatty acids, better known as fish oil, can help lower eye pressure and improve circulation to the eye by naturally thinning your blood. Fish oil is available in convenient, tasteless capsules from a number of manufacturers. If you don't like the occasional "fish breath" that can occur with these capsules, try flaxseed oil instead. Omega-3 fish oil or flaxseed oil is generally very safe in doses up to 2 g a day. *However, if you are taking any sort of blood-thinning medication, do not take omega-3.* Diabetics need to be cautious about taking omega-3 capsules; discuss taking these supplements with your doctor before you try them.

If you have pre-glaucoma symptoms (high ocular pressure, for example) or glaucoma, *take your medication exactly when and how your doctor prescribes!* Daily nutritional supplementation in addition to your medication could help keep glaucoma from developing or slow its progress, but discuss any supplements with your doctor.

GLUTATHIONE AND DIABETIC RETINOPATHY

People with diabetes are at greater risk for eye problems. In particular, diabetics are prone to a problem called diabetic retinopathy. The tiny blood vessels that nourish the retina at the back of your eye become constricted and develop bulges. The bulges often burst, causing little hemorrhages that kill nearby retinal cells. As the disease progresses, the macula (the central, most sensitive portion of the retina) is damaged, causing blurred vision. Eventually, the blood vessels to the retina are so damaged that they can no longer carry nutrients. At that point, abnormal blood vessels can start to form on the retina, a condition called neovascularization. These blood vessels are very fragile and often break, leaking blood into the vitreous humor (the jelly-like substance in the eyeball). The leakage can lead to temporary dimness in your vision. It also causes scars on your retina, which can eventually lead to permanent vision loss.

The longer you are diabetic and the more severe or uncontrolled your diabetes is, the more likely you are to develop diabetic retinopathy. To reduce your lifetime chances of diabetic eye problems, it is essential to keep your blood sugar under control through careful attention to your diet, exercise, and medication. If you also have high blood pressure, it is vital to keep that under control as well.

Even if you follow your diabetes diet faithfully, however, you could be missing some vital nutrients that might help slow or even prevent diabetic retinopathy. I strongly recommend an outstanding book by Dr. Julian Whitaker called *Reversing Diabetes* (Warner Books, 1987). Dr. Whitaker recommends a program of nutritional supplements that includes chromium, magnesium, beta carotene, vitamin B_6, and omega-3 fatty acid. Magnesium supplementation is very important, particularly if you have trouble controlling your blood sugar levels. Several studies have shown that diabetics with low magnesium levels are the ones most likely to develop not only retinal disease but cardiovascular problems as well. Other studies show that diabetics with severe retinopathy are significantly deficient in magnesium. Even if you are not convinced of the value of the other supplements I recommend here, I urge you to discuss magnesium supplements with your doctor.

To my diabetic patients, I usually suggest Dr. Whitaker's basic program along with additional antioxidants and cofactors. It's vital to maintain high levels of glutathione and taurine in your eyes to protect your retina and fight against the cell wall destruction caused by diabetic retinopathy.

To keep your glutathione level high, you need to add NAC and the cofactors selenium and zinc to your daily supplements. Even though it's important to keep your taurine levels high as well, I don't recommend taking taurine supplements if you have diabetes or blood sugar problems of any sort. Taurine increases some of the effects of insulin, which could be dangerous. In general, diabetics should consider taking the basic supplements for healthy eyes suggested on page 141, along with 3,000 to 10,000 mg a day of anthocyanoside extract.

Supplements are helpful in addition to, not instead of, other steps to control your diabetes and high blood pressure. Discuss adding supplements such as chromium picolinate with your doctor before you try them.

UVEITIS, IRITIS, AND AMINO ACIDS

Uveitis is a general name for any inflammation of the uveal tract in your eye, which includes the choroid, the ciliary body, and the iris. The choroid is a layer of tissue that lines the inside of your eye. Rich in blood vessels, the choroid supplies nutrients and oxygen to your eye. Toward the front of your eye, the choroid thickens to form the ciliary body. Your iris (the colored portion of your eye) is formed by muscles attached to the ciliary body. The muscles of your iris regulate how much light enters your eye by widening or contracting your pupil. The ciliary body also holds the lens of your eye in place.

Iritis is an inflammation of the iris (and sometimes the ciliary body and even the choroid). Symptoms of iritis include discomfort or pain in the eye, redness, sensitivity to light, and often a slight reduction in vision. The cause is unknown, although doctors suspect that in many cases a virus is to blame; other causes could include toxins and allergens. Ophthalmologists generally treat iritis with a nonsteroidal anti-inflammatory drug (NSAID) in the form of eye drops. This often clears up the trouble, but some patients don't respond well. Others may have fairly frequent recurrences. Chronic iritis can later lead to glaucoma or cataracts.

Whenever you have an inflammation, your body produces extra free radicals, and your eye is no exception. Since your eyes produce large numbers of free radicals even when they are healthy, iritis or uveitis means that you need more antioxidants to mop up those extra free radicals before they damage your cells.

If the inflammation affects your choroid, you need extra antioxidants even more. It's the job of the many tiny blood vessels in your choroid to bring the vitamin C and the amino acids needed to synthesize glutathione to the eye. If your choroid itself is inflamed, it becomes inefficient, and your eye will be starved for nutrients. Help it out by providing extra building blocks.

If you have iritis or uveitis symptoms, even mild ones, see your ophthalmologist at once. Follow his or her directions for treating the problem. In addition, I suggest you take the supplements for healthy eyes recommended on page 141, to prevent recurrences, keep taking them after your eyes clear up.

HERPES EYE INFECTIONS

The herpes simplex virus (type I) is the virus that causes those annoying and painful cold sores. Sometimes the herpes simplex virus attacks your cornea (the thin, clear membrane that covers your iris and pupil) and causes a painful ulcer. Corneal herpes symptoms include pain and redness in the sclera (white) of the eye. If you have a herpes ulcer, you probably won't be able to see it. (If you can see a whitish patch on the cornea, it is an ulcer, but one that has been caused by something else, such as a foreign object in your eye. See your doctor at once.)

Herpes corneal ulcers need prompt medical treatment. If the ulcer is untreated, it could scar your cornea, which leads to vision loss. If the scarring is bad enough, you might need a corneal transplant operation. Fortunately, herpes ulcers usually respond well to medication.

The real problem with herpes corneal ulcers is that they tend to recur. Not only is this painful, it's inconvenient, since you'll probably miss some days of work and have to make a number of visits to the ophthalmologist before it clears up again. And there's always the chance of corneal scarring.

Herpes recurs because the virus never really goes away—it just lies dormant until something triggers it again. For many herpes patients, ultraviolet light is a trigger. Avoid UV light by wearing 100 percent UV-blocking sunglasses (see Figure 7.1) when you are outside in daylight.

Another possible herpes trigger is diet. Many herpes patients say that foods that are high in the amino acid arginine, such as nuts, chocolate, and gelatin, trigger their symptoms. On the other hand, foods that are rich in the amino acid lysine seem to block herpes. Lysine-rich foods such as brewer's yeast, milk, meat, and soybeans could help prevent recurrences of herpes corneal ulcers. Taking lysine supplements could also help. If you have herpes, avoid taking protein powders or general amino acid supplements, since these will contain arginine.

Many doctors suggest taking daily supplements of lysine, vitamin C, and zinc to help prevent herpes flare-ups. If you have corneal herpes, taking additional antioxidants and their cofactors may help reduce the severity of the attacks and prevent recurrences. Take the basic healthy eyes formula listed on page 141. You might also want to try 2,000 to 3,000 mg of lysine.

REFERENCES

American Diabetes Association, "Magnesium Supplementation in the Treatment of Diabetes," *Diabetes Care*, vol. 13, pp. 1065–67, 1992.

Andreanos, D. "Normal-Pressure Glaucoma Linked to Poor Blood Flow," *Ophthalmology Times*, p. 8, October 15, 1991.

Asregadoo, E. R. "Blood Thiamine Levels and Ascorbic Acid in Chronic Open Angle Glaucoma," *Annals of Ophthalmology*, vol. 11, pp. 1095–100, 1976.

Christen, William. "The Use of Vitamin Supplements and the Risk of Cataracts Among U.S. Male Physicians," *Journal of Public Health*, vol. 84, no. 5, pp. 788–92, May 1994.

Costagliola, L. T., et al. "Cataract Risk Factors," *Metabolic, Pediatric, and Systemic Ophthalmology,* vol. 14, pp. 31–36, 1991.

Cruickshanks, K. J., R. Klein, and B. E. K. Klein. "Sunlight and Age-Related Macular Degeneration," *Archives of Ophthalmology,* vol. 111, pp. 514–18, 1993.

Dezardi, X. "Reduced Glutathione (GSH) Concentration Is Low and Related to the Severity of the Disease in Human Cataract," *Investigative Ophthalmology,* vol. 33: ARVO Abstracts 1078, March 15, 1993.

Drews, C. D., et al. "Dietary Antioxidants and Age-Related Macular Degeneration," *Investigative Ophthalmology,* vol. 34: ARVO Abstracts 2237, March 15, 1993.

Ferrer, J. V., et al. "Senile Cataract: A Review of Free-Radical Related Pathogenesis and Antioxidant Prevention," *Archives of Gerontology,* vol. 13, pp. 51–59, 1991.

Handelman, G. J., and E. A. Dratz. "The Role of Antioxidants in the Retina," *Advances in Free Radical Biology and Medicine,* vol. 2, pp. 1–89, 1986.

Hoeve, J. van de. "Eye Lesions Produced by Light Rich in Ultraviolet Rays," *American Journal of Ophthalmology,* vol. 3, pp. 178–94, 1920.

Jacques, P. F., and L. T. Chylack Jr. "Epidemiologic Evidence of a Role for the Antioxidant Vitamins and Carotenoids in Cataract Prevention," *American Journal of Clinical Nutrition,* vol. 53, pp. 352S–55S, 1991.

Lampson, W. G., et al. "Potentiation of the Actions of Insulin by Taurine," *Canadian Journal of Physiology and Pharmacology,* vol. 61, pp. 457–63, 1983.

McNair, P., et al. "Hypomagnesia, a Risk Factor in Diabetic Retinopathy," *Diabetes,* vol. 27, pp. 1075–77, 1978.

Mehra, K. S., and P. K. Shukla. "Vitamin A and Primary Glaucoma," *Glaucoma,* vol. 4, pp. 226–27, 1982.

Nutrition Reviews, "Taurine Function Revealed by Its Nutritional Requirements in the Kitten," vol. 37, pp. 121–23, 1979.

Robertson, J. McD., A. P. Donner, and J. R. Trevithick. "A Possible Role for Vitamins C and E in Cataract Prevention," *American Journal of Clinical Nutrition,* vol. 53, pp. 346S–51S, 1991.

Sakai, T., M. Murata, and T. Amemiya. "Effect of Long-Term Treatment of Glaucoma with Vitamin B_{12}," *Glaucoma,* vol. 14, pp. 167–70, 1992.

Sardi, Bill. *Nutrition and the Eyes* (Montclair, Cal.: Health Spectrum Publishers, 1994).

Seddon, Johanna M., et al. "Dietary Carotenoids, Vitamins A, C, and E, and Advanced Age-Related Macular Degeneration," *Journal of the American Medical Association,* vol. 272, pp. 1413–20, November 9, 1994.

Sternberg, P. "Protection of Retinal Pigment Epithelium from Oxidative Injury by Glutathione and Precursors," *Investigative Ophthalmology*, vol. 34, pp. 3661–68, 1993.

Varna, S. D. "Scientific Basis for Medical Therapy of Cataracts by Antioxidants," *American Journal of Clinical Nutrition*, vol. 53, pp. 335S–45S, 1991.

Virno, M., et al. "Oral Treatment of Glaucoma with Vitamin C," *Eye, Ear, Nose, and Throat Monthly*, vol. 46, pp. 1502–8, 1967.

Weale, R. "Why Does the Human Visual System Age in the Way It Does?" *Experimental Eye Research*, pp. 49–55, January 1995.

West, S. K., et al. "Exposure to Sunlight and Other Risk Factors for Age-Related Macular Degeneration," *Archives of Ophthalmology*, vol. 107, pp. 875–79, 1989.

Whitaker, Julian. *Reversing Diabetes* (New York: Warner Books, 1987).

CHAPTER 8

GLUTATHIONE AND OTHER AILMENTS

Glutathione can play an important role in treating or even preventing a number of serious ailments. A number of other medical problems respond well to glutathione treatment in conjunction with standard medical treatment. Lung problems such as asthma, skin problems such as psoriasis, and intestinal problems such as ulcers may all benefit from glutathione supplementation. In addition, it may help heal wounds and infections and aid in recovering from surgery.

A dramatic example of glutathione's ability to help a chronic disease is shown by the case of a patient named Wilson. Since childhood, Wilson suffered from severe asthma. He missed school a lot as a child; as an adult, he missed many days of work. He was taking an array of powerful medications, but his asthma was still just barely under control. Someone—not his doctor—suggested that Wilson see a nutritionist. He called me, and we set up an initial visit. In that consultation, I was amazed to learn that none of Wilson's many doctors over the years had *ever* discussed the role of nutrition in managing asthma.

I arranged for Wilson to be tested for food allergies. Not surprisingly, he was allergic to a number of foods he commonly ate. We modified his diet to avoid his trigger foods and put him on a program of antioxidant supplements that included vitamins A and C, magnesium, amino acids, and glutathione. Within three months, Wilson's life changed. His asthma attacks were less frequent and less severe. His doctor was able to reduce his medicine dosages by half, which also cut in half their unpleasant side effects. Wilson continues to work with me and his doctor, and his asthma continues to improve.

The combined medical and nutritional approach that helped Wilson may also help you. Glutathione, nutrition, and a holistic medical approach have helped many of my patients leave their debilitating illnesses behind and return to an active, healthy life.

ASTHMA AND OTHER LUNG PROBLEMS

As Wilson's case shows, glutathione, nutrition, and medicine can combine to excellent effect in the treatment of asthma. To understand why, you need to know what asthma is and what triggers an attack. A chronic inflammatory condition of the lungs, asthma causes periodic attacks of wheezing and difficulty in breathing. The airways in the lungs of people with asthma are unusually sensitive to allergens (including some foods) and irritants (such as tobacco, smoke, air pollution, or chalk dust) in the air; emotional stress, exercise, and even cold weather can also cause attacks. Asthma symptoms begin when these triggers cause the linings of the airways to swell up; this narrows the airways and makes it hard for the person to breathe. The muscles that surround the airways can then go into spasms, which makes breathing even harder. The inflamed linings of the airways produce mucus, which clogs up the airways and makes breathing difficult. In addition to having difficulty breathing, asthma sufferers feel a painless tightening in the chest and wheeze, sometimes very noticeably, when they breathe.

Asthma is quite common—nearly ten million Americans suffer from it, including three million children. Symptoms are often mild and respond well to treatment, but the disease must always be taken very seriously. Even a mild asthma attack is uncomfortable and distressing; a severe one can be life-threatening. Someone who has mild asthma can still have an unexpectedly severe attack that needs emergency medical treatment. If you suspect asthma in yourself or your child, see your doctor. If you or your child have asthma and the attacks are getting more frequent or more severe, see your doctor at once.

If you have an asthma attack, immediately take the medication your doctor has prescribed. In most cases such medication is very effective and will halt an attack quickly. If the attack is unusually severe for you or if the medication doesn't help, get medical assistance at once.

Two different groups of factors can trigger asthma. Extrinsic (atopic) asthma is generally caused by allergens such as pollen or animal dander that trigger an allergic response. Intrinsic asthma is generally a nonallergic response (the immune system is not involved) caused by factors such as air pollution, stress, respiratory infection, cold weather, or exercise.

Since many asthmatics have allergies to pollen or foods, it's important to work with your doctor to discover your allergies and find ways to reduce your exposure. Air conditioners, for example, can reduce the amount of pollen and other airborne allergens in the home (be sure to clean the filters regularly). Some asthmatics have attacks because the trigger is a food; the most common culprits are milk, eggs, seafood, chocolate, citrus, wheat, and nuts and peanuts. If you know that your asthma is triggered by these foods, avoid them.

About 5 percent of asthmatics are sensitive to the salicylates found in certain medications, such as aspirin and other nonsteroidal anti-inflammatory drugs (NSAIDs), and to sulfites, which are widely used as preservatives in foods and beverages. A number of foods naturally contain salicylates and should be avoided if you are sensitive. Tea, root beer, corned beef, avocados, cucumbers, green peppers, olives, potatoes, tomatoes, and some fruits, particularly apples, berries, cherries, grapes, melons, peaches, and plums, are all on the salicylate list. Sulfites are commonly used in restaurants and convenience foods to help preserve the freshness of shrimp, potatoes, dehydrated soups, and other foods. They are also found naturally in beer, wine, and dried fruits (especially apricots). Monosodium glutamate (MSG), another widely used food preservative, can also trigger asthma attacks in sensitive individuals. Asthmatics who are sensitive to sulfites or MSG should be very careful to avoid these ingredients.

Given the above, it's not surprising that many of my asthmatic patients have been able to reduce the frequency and sometimes the severity of their attacks simply by making some dietary changes. Recent research into the role of glutathione in treating inflammatory lung diseases suggests, however, that glutathione could have an important role in helping asthmatics. The reason: free radicals.

Asthmatics may be more susceptible to cell damage from free radicals—either from outside sources such as cigarette smoke or air pollution or from internal sources such as the inflammatory response of an allergic reaction to pollen. In other words, they are more likely to be in a state of oxidative stress. But why are they more susceptible? In some cases, they are simply genetically predisposed to this sort of reaction. Their bodies may naturally produce less of a particular enzyme that helps cells, especially red blood cells, absorb glutathione. And because their cells contain less glutathione, they are more vulnerable to damage from free radicals.

It's also possible that heavy metal poisoning, particularly from lead and copper, can block the pathways that help cells absorb glutathione. High levels of iron may also be a factor. Environmental factors such as this could help explain why asthma is so shamefully prevalent in poor neighborhoods, particularly among children. Poor housing often contains rusting iron water pipes and old lead paint; such housing is often located near incinerators, old industrial areas, and waste dumps. People in these circumstances are exposed to excess amounts of environmental toxins and cell-damaging particulates.

Because free radicals and environmental toxins can play a role in asthma, it makes sense to discuss supplements of antioxidants and glutathione with your health care provider. In particular, studies suggest that supplemental selenium may help increase glutathione uptake by the cells and could help prevent asthma attacks. Vitamin C has also been clinically proven to help treat asthma symptoms, particularly if the symptoms are caused by free radicals from exposure to air pollution. Supplemental glutathione may also be especially helpful for asthmatics, since their tissue levels of glutathione tend to be low.

Supplemental cysteine in the form of NAC may help asthmatics by thinning the mucus in the bronchial tubes and lungs and helping to prevent respiratory infections. NAC in aerosol form, administered with an inhaler, is sometimes prescribed for asthmatics. In supplement form, NAC is less effective but can still be very helpful. If you'd like to try aerosol or supplemental NAC, discuss it with your doctor.

Daily Supplements for Asthma

If you have asthma, avoid your known triggers and take your medicine as your doctor prescribes, but also consider adding these supplements:

Vitamin A	10,000 IU
Vitamin C	3,000 mg
Vitamin E	400 mg
Mixed carotenes	2,500 IU
Selenium	100 mcg
Magnesium	800 mg
GSH	250 mg
NAC	1,000 mg
Lipoic acid	400 mg

For best results, divide the doses into morning and evening halves. In addition, I recommend a good daily multivitamin from a reputable manufacturer such as Solgar or Nature's Way. Be sure it provides all the B vitamins.

Asthmatics can take some additional steps to restore their oxidative balance and clear their bodies of the toxins that may be contributing to their attacks. It's possible that a leaky gut is allowing undigested food particles to enter your blood from your small intestine; the particles then may trigger an allergic response that leads to an asthma attack. If you think leaky gut syndrome is contributing to your asthma, look again at the discussion in chapter 2. Work with your health care provider to diagnose and treat the problem. It's possible that prescription or nonprescription medicines you take to control your asthma could be contributing to leaky gut syndrome. Even if you think this is the case, *do not stop taking your asthma medicine.* Always discuss changing medications with your doctor before you try it.

Oxidative stress, an overload of toxins, or heavy metal poisoning—or all three—could also be contributing to your asthma. Dietary changes that add plenty of fresh fruits and vegetables and whole grains to your diet can help reduce oxidative stress (avoid your food asthma triggers), but they may not be enough. Supplements of antioxidant vitamins, glutathione, and glutathione cofactors, as discussed above, may also be needed. Many of my asthmatic patients have benefited from following a metabolic cleansing program as described in chapter 2, and then going on a careful diet that maximizes their intake of foods rich in glutathione and antioxidants but avoids their trigger foods. The metabolic cleansing helps rid their system of environmental toxins and helps restore their oxidative balance.

If you suspect that heavy metal poisoning or an overload of iron is contributing to your asthma, refer to chapter 3 on environmental illness. You can do a great deal to reduce your exposure and remove these dangerous substances from your system.

Bronchitis is another lung problem that may be helped quite a bit by supplemental glutathione. Bronchitis occurs when the sensitive mucus membranes lining the airways (bronchi) leading to your lungs become inflamed. Symptoms of bronchitis include a deep cough that brings up yellowish or grayish phlegm from your lungs, wheezing, fever, and breathlessness.

Acute bronchitis is usually the result of a viral respiratory infection such as a bad cold or the flu. In general, you'll feel better within a few days, especially if you stay in bed and drink plenty of liquids. If your bronchitis symptoms don't clear up within forty-eight hours, however, call your doctor.

Chronic bronchitis—symptoms that persist or get worse—can be a serious medical condition. It usually begins with morning coughing that goes away during the day. If you smoke, you may think of this as smoker's cough. Over a period of years, chronic bronchitis can get worse. You may get an attack of acute bronchitis every time you have a minor respiratory ailment or when the weather is cold and damp. Eventually, you may have a constant cough and severe wheezing and breathlessness. People with chronic bronchitis are much more susceptible to lung infections that can lead to severe illness and pneumonia. They are also more likely to develop emphysema, heart failure, and other life-threatening medical problems.

Although air pollution can be a cause of chronic bronchitis, the chief cause is cigarette smoking. If you smoke, stop. Giving up cigarettes will help your chronic bronchitis, but you will probably have the condition for life. There is now some solid evidence that supplemental glutathione and cysteine may help reduce lung damage and could also reduce the frequency and severity of lung infections in people with chronic bronchitis.

The delicate alveoli, or tiny air sacs, that make up your lung tissue are very vulnerable to oxidative damage. Cigarette smoke and air pollution destroy the cells that make up the surface of your lungs. Supplemental GSH or NAC along with vitamin C may provide some additional antioxidant protection against this sort of lung damage and prevent the onset of chronic bronchitis. If you are a smoker or are heavily exposed to air pollution on a daily basis, I suggest taking two tablets daily of GSH 250 Master Formula along with 2,000 mg of additional vitamin C; divide the supplements into morning and evening doses. No amount of supplements, however, will give the benefits that giving up cigarettes will.

The oxidative damage that harms your alveoli also reduces the ability of macrophages in your lungs to seek out and destroy any harmful bacteria you inhale. Here too supplements of GSH or NAC and antioxidant vitamins and cofactors may help boost your ability to fight off lung infections. This is particularly important for people with chronic bronchitis, whose mucus-clogged lungs are more susceptible to infection.

Supplemental NAC can be very helpful for people suffering from chronic bronchitis. As mentioned in the discussion of asthma above, NAC can help thin the mucus in the bronchi and lungs. Thinner mucus causes less clogging of the bronchial passages, is easier to cough up, and is less hospitable to pathogens. Studies in Europe have shown that patients who take NAC on a regular basis miss significantly fewer days of work because of their chronic bronchitis. These patients also have fewer exacerbations (periods of more severe symptoms) caused by infections, perhaps because their macrophage response is improved. If you have chronic bronchitis, discuss NAC supplements with your doctor. Taking 2,000 mg a day, divided into morning and evening doses, along with supplemental zinc (15 to 30 milligrams) could help your symptoms.

GLUTATHIONE AND SKIN PROBLEMS

Glutathione and other antioxidants can sometimes be very helpful for skin, hair, and nail conditions. It may even help prevent skin cancer due to sun exposure.

Psoriasis, a persistent skin disease that causes inflammation, scaling, and itchiness, affects millions of people. Although the underlying causes of psoriasis are still poorly understood, many researchers now feel that free radicals may cause an imbalance in the pathways that control cell division. This causes the skin cells to multiply much faster than they can be shed normally. Researchers also believe that some cases of psoriasis may be a form of autoimmune disease.

Whatever the mechanism, taking glutathione and antioxidant vitamins seems to help some of my psoriasis patients. I usually recommend taking two tablets daily of the GSH 250 Master Formula, along with 1,000 mg of vitamin C, 400 IU of vitamin E, and 25,000 IU of mixed carotenes.

A sluggish liver is sometimes the culprit behind psoriasis. Many of my patients benefit from following the metabolic cleansing program discussed in chapter 2. When they have completed the program, many continue to benefit from a diet that contains no yeast or dairy products. I also usually recommend daily doses of fish oil or flaxseed oil, which both contain the essential fatty acid EPA, along with a daily dose of borage seed oil, which contains the essential fatty acid GLA.

If you are taking any prescription medications for psoriasis, discuss taking supplements with your doctor before you try them.

For unknown reasons, teenagers with acne often have low levels of zinc and glutathione in their blood. Many teens I have treated responded well to improved nutrition and supplemental zinc, selenium, and glutathione—their acne noticeably decreased in severity. If acne is a problem for you, try taking four Glutathione

250 Master Formula capsules daily. Avoid fats and sugar and eat a good, nutritious diet that is very high in fresh fruits, vegetables, and whole grains.

The appearance of your skin can often be helped by taking glutathione or cysteine supplements. Dry, itchy skin, for example, can easily be helped. Your prostaglandins, hormone-like substances that regulate many parts of your body's metabolism, play a role in how much moisture your skin retains. If your prostaglandins are damaged by free radicals, your skin won't retain moisture well. Adding glutathione to your system counteracts the damaging free radicals caused by sun exposure and lets the prostaglandins do their work of keeping your skin moist and supple.

Glutathione and cysteine may also help you avoid wrinkles. The fibrous connective tissue that underlies your skin is made from a protein called collagen. Cysteine makes up about one quarter of the many amino acids that link together in a spiral to form collagen. The sulfur in the cysteine provides a crucial bond to hold the collagen firmly together. Free radicals attack the collagen and break the bonds. When the underlying support for the skin is damaged, wrinkles result. By making sure you have plenty of antioxidant glutathione to quench the free radicals, and plenty of sulfur-containing cysteine to repair any damage that may occur, you could help prevent premature wrinkles.

About 10 percent of your keratin, the horny layer of the skin that forms your nails and hair, is made up of cystine, the oxidized form of cysteine. Cystine forms a double sulfur bond that links molecules to form a tough protein. If you're not getting enough sulfur-containing cysteine, you won't build strong cystine links in your keratin. The shortage could make you develop brittle nails and hair or even start to lose your hair. If you are losing your hair abnormally (not as part of hereditary baldness or from medical treatment), 1,000 mg of cysteine in the form of NAC daily may halt the loss.

To maintain your skin and hair at their best, I recommend taking two Glutaplex tablets and a good multivitamin daily to be sure you're getting enough antioxidants and cofactors. For moist, supple skin, stay well hydrated by drinking six to eight 8-ounce glasses of liquid a day. Pure water is simple, inexpensive, and non-caloric, but diluted fruit juices, mild herbal teas, and the like are all acceptable. To protect yourself against the damaging ultraviolet rays in sunlight, wear sunscreen with a high skin protection factor (SPF) and wear UV-blocking sunglasses (see Figure 7.1).

HELP FOR DIGESTIVE PROBLEMS

Glutathione and glutamine, a nonessential amino acid, can be helpful for digestive problems, especially ulcers and diarrhea.

Until quite recently, doctors ascribed most ulcers—painful, craterlike sores in the lining of the stomach, esophagus, or upper part of the small intestine—to excess stomach acid caused by stress and worry. In addition to treating the ulcer with antacids and a bland diet, they often prescribed tranquilizers. Recently, however, researchers have shown most ulcers are caused by a bacterium called *Helicobacter pylori*. In fact, somewhere between 75 and 90 percent of all duodenal ulcers (in the upper part of the small intestine, or duodenum) are caused by *H. pylori*. This slow-growing, corkscrew-shaped organism causes an ulcer by burrowing under the protective mucus lining of your stomach and attaching to the cells underneath. This weakens the mucus lining, allowing powerful stomach acid and a digestive enzyme called pepsin to come in contact with the cells under the lining and irritate them, causing an ulcer. In addition, ammonia and other toxic substances secreted by the bacteria itself cause inflammation of the nearby cells. *H. pylori* infection can sometimes lead to stomach cancer.

In some cases, ulcers are caused by powerful drugs taken to treat other conditions. Prescription and over-the-counter nonsteroidal anti-inflammatory drugs (NSAIDs) such as aspirin or ibuprofen, which are often prescribed to treat arthritis, can cause stomach discomfort or an ulcer if taken for long periods. If you take these drugs and start to have uncomfortable ulcer symptoms such as gnawing or burning pain in the abdomen, very severe heartburn, nausea, vomiting, or loss of appetite, call your doctor at once.

Today, treatment for ulcers caused by *H. pylori* usually consists of antibiotics, sometimes along with pink bismuth (Pepto-Bismol). The ulcer usually clears up permanently within a few weeks. Ulcers caused by drug treatment often go away once the treatment stops.

While you are on the drug, try taking supplements of the amino acid glutamine to help prevent gastritis and ulcers by helping to maintain a healthy layer of thick mucus in the linings of the stomach and small intestine. Taking 1,000 mg of supplemental glutamine thirty minutes before taking the drug has been shown to be very helpful for some patients. If you must stay on an NSAID for a long period, discuss the possibility of ulcers with your doctor. Some prescription medications can help prevent drug-induced ulcers, but glutamine may be preferred as a more natural approach that is less likely to cause side effects. If your ulcer persists after you have stopped taking the drug, *H. pylori* might be the real culprit.

If you have a drug-related duodenal ulcer, glutamine may help it heal faster. Glutamine nourishes the cells that line the small intestine and helps them repair themselves.

Whether your ulcer was caused by bacteria or drugs, smoking will make it worse and slow the healing process. Again, if you smoke, stop.

How can glutathione help your ulcer? Actually, this is a rare case where supplemental glutathione may not be a good idea. Some researchers believe that the *H. pylori* bacteria survive by eating the macrophages and neutrophils that arrive to cope with the inflammation. Since glutathione can improve macrophage activity, you might actually worsen the problem with supplements. As discussed above in the section on respiratory problems, supplemental cysteine helps to thin mucus, which is undesirable if you have an ulcer. On the other hand, any inflammation in your body means a higher level of free radical production and hence a greater need for glutathione's vital antioxidant activity. To be on the safe side with *H. pylori* infection, avoid supplemental glutathione and cysteine. Try adding supplements of antioxidant vitamins, including vitamin E and mixed carotenes. Avoid high doses of vitamin C; it is acidic and may aggravate your ulcer.

If your ulcer is not bacterial, however, glutathione may indeed help reduce the inflammation and fend off the additional free radicals. To avoid thinning the mucus lining of your stomach and small intestine, don't take cysteine supplements.

Supplemental glutamine may help heal all ulcers. To understand why, let's go back to the basics of amino acids. Glutathione is made up of three amino acids: cysteine, glycine, and glutamate. Glutamate and glutamine are actually forms of the nonessential amino acid glutamic acid. Since only very small amounts of glutamine are found naturally in foods, and while glutamic acid is abundant (especially in plant foods), your body readily converts glutamic acid to make most of the glutamine it needs. Glutamine plays a role in a wide variety of important metabolic functions; in fact, it is the most common amino acid in the human body. Glutamine is essential for the synthesis of DNA, necessary for cell division and cell growth. You need glutamine to heal wounds and make repairs to your body. Glutamine is the energy source for your immune system and also nourishes and heals your gastrointestinal tract.

If your body needs large amounts of glutamine—if you are recovering from a serious injury or burn, for example—you may not be able to get enough from your diet. Your body will then convert all its available glutamic acid to glutamine—to the point of breaking down muscles cells to get it—leaving very little for synthesizing glutathione. A shortage of glutathione leads to oxidative stress and free radical damage to your cells; toxins normally carried away by glutathione may accumulate in your tissues instead. Recent nutritional research has shown that patients who received supplemental glutamine after surgery recovered faster with fewer complications.

As discussed above, glutamine may help prevent ulcers caused by powerful drugs. If you do get an ulcer, glutamine can help heal it by nourishing the cells of your stomach and small intestine, providing fuel for your immune system, and helping your cells repair the damage. Glutamine supplements, usually in the form of 500 mg capsules or tablets, are readily available at well-stocked health food stores.

Glutamine may also help treat serious diarrhea caused by AIDS or by other intestinal problems such as ulcerative colitis. Fairly large doses of glutamine—as high as 40,000 mg—may be needed. The glutamine improves the absorption of water through the colon, which helps to relieve the diarrhea. *Dosages that high should be taken under supervision, and only by those with diarrhea caused by a serious medical problem. Don't treat minor diarrhea from indigestion or a twenty-four-hour stomach virus with glutamine.*

AMINO ACIDS AND
CHRONIC FATIGUE SYNDROME

Chronic fatigue syndrome (CFS) is an illness that causes severe and disabling fatigue, along with a wide array of other symptoms that may include headaches, muscle and joint pain, inability to concentrate, depression, sore throat, and tender lymph nodes under the armpit. CFS is a controversial disease in part because its origin is still unknown. Some health care providers dismiss it as "yuppie flu" or ascribe the symptoms to depression or stress. Even the number of people who have CFS is controversial. According to the Federal Centers for Disease Control and Prevention, the syndrome affects 100,000 to 250,000 people nationwide. Other researchers point out that this group includes only those who have sought medical help; it is possible that the number is as high as 1.25 million.

The criteria for diagnosing CFS have recently been more sharply defined by medical researchers. The new guidelines reduce the number of symptoms a patient must have to be diagnosed with CFS. The most important symptom is unexplained, persistent, or relapsing chronic fatigue that has no other cause and is not relieved by rest.

Just as no one knows what causes CFS, no one knows exactly how to treat it. Most health care providers recommend getting lots of rest and eating a good, nutritious diet with plenty of whole grains and fresh fruits and vegetables. Some of my patients with CFS have benefited from a metabolic cleansing program as discussed in chapter 2, but most find amino acid supplements more helpful. For unknown reasons, many CFS patients have low blood levels of essential amino acids. They tend to be particularly low in tryptophan, phenylalanine, taurine, isoleucine, and leucine. Since you need the essential amino acids to produce adenosine triphosphate (ATP), the basic fuel for your cells, a shortage could interfere with your ability to use ATP for energy. CFS patients also often have low levels of the mineral magnesium, necessary for the utilization of ATP.

Daily Supplements for CFS

To compensate for their low amino acids, I recommend that my CFS patients take a large daily dose of 10 to 15 g of free-form amino acid powder. I also recommend daily intake of at least the RDA for magnesium, about 350 mg. Too much magnesium can cause diarrhea, so it is probably best to get this mineral by taking a high-quality daily multivitamin/mineral supplement and eating foods rich in magnesium, such as peanuts, bananas, avocados, cashews, wheat germ, and milk. One banana, for example, has 63 mg of magnesium; two tablespoons of wheat germ contain 40 mg.

Within a few weeks of starting amino acid treatment, most of my CFS patients report a significant improvement in their symptoms, especially their ability to concentrate. After three months of supplementation, most have fully recovered. One of my patients, a thirty-two-year-old executive secretary named Pamela, recovered dramatically. Pamela had become increasingly tired over a four-year period. Getting up in the morning was an effort. By early afternoon she felt "foggy" and exhausted; her joints and muscles ached. She would drag herself home from work and be in bed by seven in the evening. Needless to say, her work and social life were suffering badly. Her doctor diagnosed CFS but couldn't suggest anything more helpful than getting a lot of rest. Pamela came to me for a nutritional approach to treating her illness. We improved her overall diet by cutting out junk food and sweets and adding lots of fresh fruits and vegetables, whole grains, and easily digested proteins such as eggs and lamb. We also added 15 g a day of amino acid powder. Within two months, Pamela's four-year illness was over.

GLUTATHIONE AND OTHER PROBLEMS

Recent research suggests that glutathione or cysteine may be helpful in treating a range of other medical problems.

As discussed extensively in chapter 2, glutathione in your liver plays an essential role in removing free radicals, metabolic wastes, and toxins from your system. If your liver is damaged or diseased, it will not work efficiently. Hepatitis, a viral infection of the liver, is a fairly common disease that causes impairment of liver function. Hepatitis is generally treated simply by rest and good nutrition.

Daily Supplements for Hepatitis

Taking GSH or NAC supplements may help relieve the symptoms and speed your recovery by protecting your liver from free radical and toxic waste damage. Silymarin, or milk thistle, supplements provide additional liver protection. Many hepatitis patients also benefit from taking fairly large doses of vitamin C (reduce the dose if you get diarrhea). I suggest these supplements if you have hepatitis:

Vitamin C	2,000 mg
GSH	500 mg
NAC	1,000 mg
Lipoic acid	500 mg
Selenium	200 mcg
Silymarin	600 mg

For best results, divide the doses into three portions spread over the day.

If your liver has been damaged by severe hepatitis or alcohol, GSH or NAC may help prevent further damage. It's possible they could even reverse cirrhosis, but much more research needs to be done in this area before this is proved.

Another interesting area of research for cysteine is in preventing kidney stones. Recent studies suggest that NAC supplements could keep people who have had kidney stones in the past from forming new stones.

REFERENCES

Beloqui, O., J. Pietro, M. Suarez, et al. "N-acetylcysteine Enhances the Response to Interferon-alpha in Chronic Hepatitis C: A Pilot Study," *Journal of Interferon Research*, vol. 13, pp. 279–82, 1993.

Blaser, Martin J. "The Bacteria Behind Ulcers," *Scientific American*, pp. 104–7, February 1996.

Bock, S. A. "Food-Related Asthma and Basic Nutrition," *Journal of Asthma*, vol. 20, pp. 377–81, 1983.

Bralley, J. Alexander, and Richard S. Lord. "Treatment of Chronic Fatigue Syndrome with Specific Amino Acid Supplementation," *Journal of Applied Nutrition*, vol. 46, no. 3, pp. 74–78, 1994.

Fan, J., and S. J. Shen. "The Role of Tamm-Horsfall Mucoprotein in Calcium Oxalate Crystallization. N-acetylcysteine—A New Therapy for Calcium Oxalate Urolithiasis," *British Journal of Urology*, vol. 74, pp. 288–93, 1994.

Freedman, B. J. "A Diet Free from Additives in the Management of Allergic Disease," *Clinical Allergy*, vol. 7, pp. 417–21, 1977.

Greene, Lawrence S. "Asthma and Oxidant Stress: Nutritional, Environmental and Genetic Risk Factors," *Journal of the American College of Nutrition*, vol. 14, no. 4, pp. 317–24, 1995.

James, M. B. "Hair Growth Benefits from Dietary Cysteine-Gelatine Supplementation," *Journal of Applied Cosmetology*, vol. 2, pp. 15–27, 1983.

Juhlin, L., L. Bedquist, G. Echman, et al. "Blood Glutathione-Peroxide Levels in Skin Diseases: Effect of Selenium and Vitamin E Treatment," *Acta Dermat. Vener.*, vol. 62, pp. 211–14, 1982.

Lemy-Debois, N., G. Frigerio, and P. Lualdi. "Oral Acetylcysteine in Bronchopulmonary Disease: Comparative Clinical Trial with Bromhexine," *European Journal of Respiratory Disease*, vol. 61, pp. 78–80, 1980.

Lindahl, O., et al. "Vegan Diet Regimen with Reduced Medication in the Treatment of Bronchial Asthma," *Journal of Asthma*, vol. 22, pp. 45–55, 1985.

Maurice, P. D. L., B. R. Allen, A. S. J. Barkley, et al. "The Effects of Dietary Supplementation with Fish Oil in Patients with Psoriasis," *British Journal of Dermatology*, vol. 117, pp. 599–606, 1987.

Morris, P. E., and G. R. Bernard. "Significance of Glutathione in Lung Disease and Implications for Therapy," *American Journal of Medical Science*, vol. 307, pp. 119–27, 1994.

Murray, Michael T. "Chronic Fatigue Syndrome," *American Journal of Natural Medicine*, vol. 2, no. 6, pp. 12–21, July/August 1995.

Rasmussen, J. B., and C. Glennow. "Reduction in Days of Illness After Long-Term Treatment with N-acetylcysteine Controlled-Release Tablets in Patients with Chronic Bronchitis," *European Respiratory Journal*, vol. 1, pp. 351–55, 1988.

Riise, G. C., S. Larsson, P. Larsson, et al. "The Intrabronchial Microbial Flora in Chronic Bronchitis Patients: A Target for N-acetylcysteine Therapy?" *European Respiratory Journal*, vol. 7, pp. 94–101, 1994.

Schwartz, J., and S. T. Weiss. "Dietary Factors and Their Relation to Respiratory Symptoms," in The Second National Health and Nutrition Examination Survey, *American Journal of Epidemiology,* vol. 132, no. 1, pp. 67–76, 1990.

Shabert, Judy, and Nancy Ehrlich. *The Ultimate Nutrient: Glutamine* (Garden City Park, N.Y.: Avery Publishing Group, 1994).

Stevenson, D. D., and R. A. Simon. "Sensitivity to Ingested Metabisulfites in Asthmatic Subjects," *Journal of Allergy and Clinical Immunology,* vol. 68, pp. 26–32, 1981.

Weber, G., and K. Galle. "The Liver, a Therapeutic Target in Dermatoses," *Med. Welt,* vol. 34, pp. 108–11, 1983.

EATING
YOUR GLUTATHIONE

Throughout this book I've discussed the essential role of glutathione, cysteine, antioxidant vitamins, and other dietary substances in maintaining good health and treating illness. Because I
believe very strongly that good nutrition is the cornerstone of
good health, I often recommend supplements of these vital nutrients to make sure you're getting enough of them. Try as we
might, in our harried modern society we can't always eat the right
foods all the time—and in some cases, we need more of certain
nutrients than we can get through diet alone.

Just taking supplement capsules, however, isn't a very good approach to your overall nutrition. Supplements are in addition to,
not instead of, eating well. There's no pill that substitutes for a
nutritious diet and good dietary habits. Your basic nutrition
should come from the foods you eat; supplements add the extra
amounts needed for optimal wellness and the treatment of specific
problems.

But which foods? In general, a healthy diet is one that is varied,
low in fat, high in fiber, rich in fresh fruits and vegetables and
whole grains, and low in sugar and salt. To that general recommendation I would add foods that are high in glutathione.

Figure 9.1

GLUTATHIONE CONTENT OF SELECTED FOODS

Food	Milligrams per 100 grams
Acorn squash	14
Asparagus	26
Avocado	31
Broccoli	8
Cantaloupe	9
Grapefruit	15
Okra	7
Orange	11
Peach	7
Potato	13
Spinach	5
Strawberries	12
Tomato	11
Watermelon	28
Zucchini	7

Source: Dean P. Jones, Ralph J. Coates, et al., "Glutathione in Foods Listed in the National Cancer Institute's Health Habits and History Food Frequency Questionnaire," *Nutrition and Cancer,* vol. 17, no. 1, pp. 57–75, 1992.

GLUTATHIONE IN THE DIET

Dietary sources of glutathione fall into two categories: foods that naturally contain glutathione or its basic building block, cysteine, and foods that stimulate glutathione production in your body.

Glutathione is found in almost all fresh fruits and vegetables in amounts ranging from about 50 to 150 mg per kilogram of wet weight. (See Figure 9.1 for the glutathione content of some common foods.) Walnuts are high in glutathione, but other nuts are not. Avocados, asparagus, raw spinach, cauliflower, okra, broccoli, tomatoes, squash, and potatoes are vegetables particularly high in glutathione. Citrus fruits, all melons, strawberries, and fresh peaches are the fruits highest in glutathione. Fresh apples, pears, and bananas are also good sources of glutathione. The amount of glutathione can vary considerably depending on portion size,

Figure 9.2
AMINO ACID CONTENT OF MILK AND EGGS

Amino acid	8 oz. lowfat milk	1 large egg (50 g)
Tryptophan	113 mg	76 mg
Threonine	362 mg	302 mg
Isoleucine	486 mg	343 mg
Leucine	786 mg	537 mg
Lysine	637 mg	452 mg
Methionine	201 mg	196 mg
Phenylalanine	388 mg	334 mg
Tyrosine	388 mg	257 mg
Valine	537 mg	383 mg
Arginine	291 mg	377 mg
Histidine	218 mg	149 mg
Cysteine	74 mg	146 mg

Source: Bowes & Church's Food Values of Portions Commonly Used.

freshness, amount of cooking, and so on. To ensure an adequate intake, try to to eat at least five servings a day of a variety of fresh fruits and vegetables. Dairy products, grain products, and prepared foods in general are low in glutathione.

The cysteine content of most foods is low compared to the other essential amino acids. (See Figure 9.2 for the amino acid content of milk and eggs.) In general, good dietary sources of cysteine include eggs, meat, poultry, fish, whole grains, dairy products, and beans. Again, to ensure an adequate intake, eat a varied diet. Figure 9.3 lists the cysteine content of some selected foods.

A number of plant foods seem to stimulate your body to produce glutathione, even though the foods themselves are relatively low in amino acids. Researchers are still exploring the compounds that stimulate glutathione production. So far, they've learned that cyanohydroxybutene (CHB), a substance found in cruciferous vegetables such as broccoli, cabbage, brussels sprouts, and cauliflower, seems to stimulate glutathione production in your body. So do two other compounds found in cruciferous vegetables: sulforaphane and iberin. Glutathione-stimulating substances are also

Figure 9.3

CYSTEINE CONTENT OF SELECTED FOODS

Food and portion	Cysteine content
Apple (medium/140 g)	4 mg
Banana (medium/114 g)	19 mg
Barley (1 cup/200 g)	438 mg
Beef hamburger patty (3½ ounces/ 100 g)	250 mg
Beef sirloin (3½ ounces/100 g)	340 mg
Black beans (1 cup/172 g)	165 mg
Broccoli (½ cup/78 g)	17 mg
Cabbage (½ cup/75 g)	6 mg
Chicken (3½ ounces/100 g)	320 mg
Chickpeas (1 cup/240 g)	161 mg
Egg noodles (1 cup/160 g)	208 mg
Flounder (3 ounces/85 g)	220 mg
Kidney beans (1 cup/256 g)	146 mg
Lentils (1 cup/198 g)	234 mg
Lowfat cottage cheese (1 cup/225 g)	287 mg
Lowfat yogurt (1 cup/225 g)	650 mg
Oatmeal (1 cup/230 g)	146 mg
Peanut butter (2 tablespoons/32 g)	100 mg
Potato (1 medium/200 g)	59 mg
Rice (1 cup/205 g)	113 mg
Spinach (½ cup/90 g)	32 mg
Tofu (½ cup/100)	139 mg
Watermelon (1 cup/160 g)	3 mg
White beans (1 cup/260 g)	207 mg
Whole-wheat flour (1 cup/120 g)	381 mg

Source: Bowes & Church's Food Values of Portions Commonly Used.

found in spinach, cantaloupe, and watermelon. The best source of all, however, may be parsley.

Used as a healing herb for centuries, parsley *(Petroselinum crispum)* today is thought of mostly as a garnish. In fact, however, you might be better off skipping the main course and eating the parsley sprig instead. Parsley is an excellent source of antioxidant vitamins. Half a cup (30 g) of chopped fresh parsley contains ⅃ IU of vitamin A and 40 mg of vitamin C, more than an equivalent amount of orange. That half cup also contains 41 mg of calcium.

Fresh parsley is easily found in the produce section of your supermarket, and it's very easy to grow at home in the garden or in a pot on the window sill. The flat-leaf kind, also called Italian parsley, is more flavorful than the traditional curly-leaved variety, but both are equally nutritious. To get the glutathione boost, use only fresh parsley.

Caution: In large amounts, parsley can act as a diuretic. Parsley contains psoralens, compounds that in large amounts could cause sun sensitivity in very fair-skinned people. In large amounts, parsley can cause uterine contractions. Pregnant women should avoid parsley in large amounts.

EGGS AND GOOD HEALTH

Because eggs contain cholesterol, many people mistakenly think that eggs are bad for you. In my experience, just the opposite is true: Eggs are a valuable source of amino acids and other important nutrients. They are highly nutritious, inexpensive, and easy to prepare. One large egg contains all the essential amino acids and many of the nonessential ones as well, including cysteine, glycine, and glutamic acid—the building blocks of glutathione. In addition, they are an excellent source of dietary sulfur, vital for synthesizing cysteine: one large egg contains 82 g of sulfur.

Eggs are such a good protein source that they are the standard by which other dietary proteins are measured. Based on the essential amino acids it provides, an egg is second only to mother's milk for human nutrition. The biological value of an egg (the rate at which its protein is used for growth) is 93.7 out of a possible 100; by contrast, beef rates only 74, soybeans 72, and whole wheat 64.

One large egg contains approximately 213 mg of cholesterol. The recommended daily cholesterol intake is currently 300 mg, but this figure is being extensively rethought by the medical and nutritional community and will probably be revised upward. Based on my experience as a nutritionist, I feel there's no need to wait to get the health benefits of eggs. If your cholesterol levels are within normal limits, you can safely eat one or two eggs a day. Remember, your body needs cholesterol to maintain and repair your cells, manufacture vitamin D, and produce many hormones and enzymes. Just as important, cholesterol helps transport glutathione through your body and into your cells.

You make most of the cholesterol you need in your liver. High levels of blood cholesterol are associated with heart disease, but eating cholesterol in foods does not necessarily raise your blood cholesterol. In fact, if you're trying to lower your cholesterol level, eggs may actually help. Saturated fat—the fat found in animal foods and tropical oils—raises LDL ("bad") cholesterol levels more than any other element in the diet. Polyunsaturated and monounsaturated fats, however, may lower cholesterol levels when they replace saturated fats in the diet. One large egg contains 4.5 grams of fat, but almost all of it is polyunsaturated and monounsaturated fat. You'd be much better off eating two large eggs (100 g) than 3.5 ounces (100 g) of steak. The eggs contain 150 calories and 5.3 grams of fat; the steak contains 304 calories and 22 grams of fat. An extra health bonus from the egg yolks: They are one of the few natural sources of vitamin D.

Because eggs contain apoproteins, which play a role in cholesterol transport, I often recommend them for patients who suffer from the neurotoxic effects of environmental illness. These patients actually need cholesterol to repair the damage to the delicate membranes of their nerve cells. Eating eggs helps these patients manufacture apoproteins, which transport the needed cholesterol to the damaged places.

GLUTATHIONE RECIPES

To help keep your glutathione level high, try adding parsley, spinach, broccoli, and other foods that stimulate glutathione production to your diet. Eating these foods raw or lightly cooked will best preserve their nutrients. To get the maximum amount of glutathione from these foods, purchase them fresh or frozen; canned vegetables have very little glutathione. Eggs should be thoroughly cooked before eating to prevent salmonella, a bacterial illness that causes food poisoning.

A steady diet of raw or steamed veggies gets a little dull, so I've given some of my favorite recipes for glutathione-rich foods below:

SOUPS

PARSLEY SOUP

6 cups chicken stock
¼ cup uncooked rice
3 eggs
2 tablespoons lemon juice
3 cups chopped parsley leaves
1 cup plain lowfat or nonfat yogurt

In a large pot over high heat, bring the chicken stock to a boil. Add the rice. Lower the heat until the stock is just simmering. Cover the pot, and cook for 15 minutes, or until the rice is tender. Reduce the heat to very low.

In a small bowl, beat together the eggs and lemon juice. Stir in ⅓ cup of the hot chicken stock, then pour the mixture back into the large pot. Whisk for 1 to 2 minutes. Remove the pot from the heat.

Stir in the parsley and the yogurt. Serve hot, or refrigerate for several hours and serve cold.

Serves 4 to 6

CHILLED PARSLEY SOUP
WITH TARRAGON

2 tablespoons unsalted butter
2 large leeks, sliced thin
1 pound new potatoes, peeled and cubed
6 cups chicken broth
3 cups chopped fresh parsley
2 cups light cream
2 tablespoons finely chopped fresh tarragon (not *dried*)
Salt
Pepper

In large, heavy pot over moderate heat, melt the butter. Add the leeks and cook, stirring often, until they are softened but not browned, about 5 minutes.

Add the potato cubes and chicken broth, and bring the mixture to a boil. Lower the heat, cover, and simmer until the potatoes are very soft, about 30 to 35 minutes. Stir in the chopped parsley, and remove from heat.

Using a food processor, puree the mixture in batches and return it to the pot or a soup tureen. Add the light cream and tarragon, and stir. Season to taste with salt and pepper.

Chill well before serving. If the soup thickens too much in the refrigerator, thin it with a bit more cream or milk.

Serves 4 to 6

CHILLED CUCUMBER AND SPINACH SOUP

2 tablespoons unsalted butter
1 large onion, chopped
4 cups chicken or vegetable stock
3 medium cucumbers, peeled, seeded, and diced
1 pound fresh spinach, coarsely chopped
One 8-oz. package cream cheese
¼ teaspoon nutmeg
¼ teaspoon paprika
2 tablespoons lemon juice
Salt
Pepper
½ cup light cream

In a large saucepan over moderate heat, melt the butter. Add the onion and cook, stirring often, until the onion is translucent, about 4 minutes.

Add the chicken or vegetable stock and the cucumber. Lower the heat, and simmer for 20 minutes. Add the spinach and cook, stirring often, until the spinach is just wilted, about 3 minutes.

Using a food processor, puree the soup in batches and return the soup to the saucepan over low heat. Cut the cream cheese into pieces and add to the mixture, stirring constantly until it has completely melted. Add the nutmeg, paprika, and lemon juice, and stir. Season to taste with salt and pepper.

Remove the soup from heat and stir in the light cream. Chill the soup for several hours before serving.

Serves 4 to 6

BROCCOLI AND POTATO SOUP

1 large head broccoli
2 leeks
1 tablespoon unsalted butter
2 garlic cloves, finely chopped
Pinch hot red pepper flakes
3 medium potatoes, peeled and diced
3½ cups chicken broth
¼ teaspoon grated nutmeg
Salt
Pepper

Cut the broccoli crown from the stem and cut into small florets. Trim away the tough ends of the stems. Peel the stem and cut it into small pieces.

Trim the leeks, removing all but about 2 inches of the green leaves. Rinse thoroughly under cold running water. Coarsely chop the leeks.

In a large saucepan over moderate heat, melt the butter. Add the leeks and cook, stirring often, until the leeks are wilted, about 4 minutes. Add the chopped garlic and red pepper flakes, and cook for 3 minutes.

Add the broccoli stems, the potatoes, and the chicken broth, and stir. Simmer until the potatoes are tender, about 25 to 30 minutes. Add the broccoli florets, and simmer until just tender, about 5 minutes. Stir in the nutmeg, and season to taste with salt and pepper.

Serves 4 to 6

CHILLED CANTALOUPE AND TOMATO SOUP

6 to 8 ripe tomatoes
1 cucumber, peeled, seeded, and diced
1 small onion, finely chopped
1 cup sour cream
2½ teaspoons salt
½ teaspoon finely chopped ginger
Pepper
4 tablespoons lemon juice
1 tablespoon grated lemon zest
1 large cantaloupe, seeded, peeled, and cubed
2 tablespoons chopped fresh basil

Bring a large pot of water to a boil. Add the tomatoes and cook until the skins start to crack, about 3 to 4 minutes. Remove the tomatoes and let them cool briefly. Peel and seed the tomatoes. Using a food processor, puree the tomatoes; there should be about 5 cups. Place the puree in a large mixing bowl.

Using a food processor, puree the cucumber and onion together, and add the mixture to the tomatoes. Add the sour cream, salt, ginger, pepper, lemon juice, and grated lemon zest. Stir well, cover, and chill for several hours.

In another bowl, combine the cantaloupe cubes and fresh basil. Toss well, cover, and chill for several hours.

To serve, divide the cantaloupe and basil mixture among the serving bowls, and top with the soup.

Serves 4 to 6

MEXICAN POTATO AND CANTALOUPE SOUP

1 pound potatoes, peeled and cubed
1 medium cantaloupe, seeded, peeled, and cubed
2 cups milk
4 tablespoons unsalted butter
2 tablespoons fresh lime juice
4 egg yolks
Salt
Pepper
Ground cinnamon, for garnish

In a large pot of boiling water, cook the cubed potatoes until tender, about 15 minutes. Drain well.

Using a food processor, puree the potatoes, cantaloupe, and 1 cup of the milk until smooth.

In a large saucepan over moderate heat, melt the butter. Add the potato mixture and the remaining 1 cup milk, and stir gently. Add the lime juice, and cook for 5 minutes.

In a small bowl, beat the egg yolks until they are lemon-colored. Stir the egg yolks into the soup, and immediately remove the saucepan from the heat. Add salt and black pepper to taste.

Chill the soup for several hours. Sprinkle each portion with a pinch of cinnamon before serving.

Serves 4 to 6

SPINACH AND POTATO SOUP

2 quarts water
2 large potatoes, peeled and diced
1 pound fresh spinach, coarsely chopped
1 cup coarsely chopped scallions
Salt
Pepper
½ cup sour cream
2 tablespoons fresh lemon juice, or as needed

In a large saucepan over medium heat, combine the water, potatoes, ½ cup of the scallions, and salt and pepper to taste. Bring to a boil, and skim the surface. Reduce the heat to very low, and simmer gently until the potatoes are tender, about 25 to 30 minutes. Skim the surface occasionally as needed.

Add the spinach and the remaining ½ cup scallions to the soup, and cook until the spinach is just wilted, about 2 minutes.

In a small mixing bowl, whisk together the sour cream and 1 cup of the hot soup. Return the mixture to the saucepan, add the lemon juice, and stir well. Simmer until the soup is very hot but not boiling, about 10 minutes. Add more lemon juice, if desired, and remove the soup from heat.

Serve hot or cold.

Serves 4 to 6

SPINACH AND YOGURT SOUP
WITH FRESH HERBS

1 tablespoon olive oil
2 large onions, sliced thin
½ cup chopped fresh parsley
¼ cup chopped fresh chives
¼ cup chopped fresh dill
1½ tablespoons dried oregano
1 teaspoon salt
½ cup uncooked lentils
1½ cups water
4 cups plain lowfat or nonfat yogurt
4 cups torn spinach leaves, stems removed

In a large pot over low heat, heat the olive oil and sauté the onions, parsley, chives, dill, oregano, and salt, stirring often, until the onions are softened, about 10 minutes.

Add the lentils and the water. Stir well, increase the heat, and bring the mixture to a boil. Reduce the heat to low, cover, and simmer until the lentils are cooked but not mushy, about 20 minutes. Remove from heat.

In a large mixing bowl, whisk the yogurt until it is smooth and creamy. Whisk in about ¾ cup of the soup broth, 1 tablespoon at a time. Whisk the yogurt mixture back into the pot. Add the spinach and cook over low heat, stirring constantly, until the spinach is bright green and tender and the soup is heated through, about 5 to 8 minutes. Do not let the soup boil or it will separate.

Serve hot or cold.

Serves 4 to 6

SIDE DISHES

MASHED POTATO PIE WITH CHEESE

5 to 6 medium potatoes
⅔ cup milk
2 tablespoons unsalted butter
Salt
Pepper
⅓ cup wheat germ
⅓ cup unseasoned bread crumbs
½ pound mozzarella cheese, grated
1 teaspoon paprika

Preheat oven to 350°F. Peel and quarter the potatoes. Place them in a large saucepan, add enough water to cover, and boil until the potatoes are tender, about 10 minutes. Drain well.

In a large mixing bowl, combine the potatoes, milk, and butter, and mash well. Add salt and pepper to taste.

In a small mixing bowl, combine the wheat germ and bread crumbs. Spread half the mixture on the bottom of a lightly greased 9×9-inch baking dish. Spread half the potato mixture evenly in the dish, and sprinkle with half the grated mozzarella. Top with the remaining potatoes and mozzarella. Sprinkle the paprika over the top.

Bake until the top is nicely browned, about 35 to 40 minutes. Let stand for 10 minutes before serving.

Serves 4

POTATO GNOCCHI

2 pounds russet potatoes
1½ to 2 cups flour
2 teaspoons salt
1 egg, lightly beaten

In a large pot of boiling water, cook the potatoes until tender, about 15 to 20 minutes. Drain well and let cool briefly. Peel the potatoes and mash them lightly.

Place the mashed potatoes on a clean work surface. Add 1½ cups of the flour, the salt, and the egg. Work the mixture into a smooth dough, kneading with your hands. Add additional flour, a little at a time, if the dough is too soft.

Lightly flour a clean work surface. Form about ⅓ of the dough into a cylinder and roll it with your hands until you form a rope about 1 inch in diameter. Using a sharp knife, cut the rope into ½-inch slices to form the gnocchi. Repeat with the remaining dough.

To cook the gnocchi, bring a large pot of water to a rapid boil. Drop the gnocchi into the water and cook for 8 to 10 minutes. Remove with a slotted spoon and drain in a colander.

Serve as desired.

Serves 6

CURRIED EGGS AND POTATOES

1 large onion, chopped
2 garlic cloves, finely chopped
1½-inch thick slice fresh ginger
3 teaspoons hot curry powder
⅔ cup plus 2 tablespoons water
2 pounds potatoes, cubed
3 tablespoons vegetable oil
3 large tomatoes, seeded and coarsely chopped
2 bay leaves
7 hard-boiled eggs, shelled and coarsely chopped

Using a food processor, puree the onion, garlic, ginger, curry powder, and the 2 tablespoons of water, and puree until smooth.

In large skillet over moderate heat, sauté the potato cubes in the oil, stirring often, until the cubes are lightly browned, about 10 minutes. Add the curry mixture and cook, stirring often, for 5 minutes.

Add the water, tomatoes, and bay leaves and stir well. Reduce the heat to low, cover, and simmer for 20 minutes. Uncover and simmer until the sauce is slightly thickened, about 10 minutes.

Add the eggs, stir gently, and cook until the eggs are heated through, about 4 to 5 minutes. Remove the bay leaves and serve immediately.

Serves 6

INDIAN SMOTHERED POTATOES

2 pounds potatoes, peeled and cut into eighths
2 teaspoons ground turmeric
1 tablespoon curry powder
2 teaspoons salt
½ teaspoon ground cumin
½ teaspoon pepper
⅓ cup plain lowfat or nonfat yogurt
3 tablespoons unsalted butter
2 bay leaves
1 teaspoon hot red pepper flakes
1 teaspoon sugar

Preheat the oven to 350°F.

In a large pot of boiling water, cook the potatoes until they are just tender, about ten minutes. Drain well.

In large mixing bowl, combine the turmeric, curry powder, salt, cumin, pepper, and yogurt. Add the potatoes and toss well until the pieces are coated with the spice mixture.

In a large, oven-safe casserole dish over moderate heat, melt the butter. Add the bay leaves, red pepper flakes, and sugar. When the sugar starts to turn light brown, add the potatoes. Stir gently, and remove from heat. Cover the casserole, and bake for 30 minutes or until nicely browned.

Serves 4 to 6

POTATO AND ZUCCHINI FRITTATA

2 medium potatoes, peeled and coarsely grated
1 medium zucchini, coarsely grated
1 small onion, chopped finely
¼ cup unseasoned bread crumbs
3 eggs, lightly beaten
Salt
Pepper
1 tablespoon unsalted butter

In a medium mixing bowl, combine the potatoes, zucchini, onion, and bread crumbs. Add the eggs, and season to taste with salt and pepper. Mix well.

In a 9-inch skillet over medium heat, melt ½ tablespoon of the butter. When the butter is very hot, add half the vegetable and egg mixture. Spread the mixture evenly in the skillet and cook until the bottom is browned and the eggs are just set, about 5 minutes.

Carefully flip the frittata over with a wide spatula. Or, slide the frittata onto a flat plate. Hold the plate in one hand and carefully invert the skillet over it. Turn the skillet over again so that the browned portion of the frittata is now on top.

Cook for another few minutes, or until the bottom is browned.

Repeat with the remaining egg mixture. Cut into wedges and serve.

Serves 4

CURRIED LENTILS WITH SPINACH

1 cup uncooked lentils
2 garlic cloves, finely chopped
1 tablespoon olive oil
½ pound fresh spinach, stems removed
One 14-ounce can whole tomatoes with liquid, chopped
2 tablespoons soy sauce
2 teaspoons curry powder
½ teaspoon cinnamon
¼ teaspoon ground nutmeg

In a saucepan over high heat, combine 3 cups water and the lentils. Bring to a boil. Reduce the heat, cover, and cook for 30 to 40 minutes, or until the lentils are tender. Drain well.

In a large skillet over moderate heat, heat the olive oil and sauté the garlic, stirring often, for 1 minute. Add the spinach, cover, and cook until the leaves are just wilted, about 3 minutes.

Add the cooked lentils, tomatoes, soy sauce, curry powder, cinnamon, and nutmeg to the skillet. Stir well. Reduce the heat, cover, and simmer for 15 minutes.

Serves 4

NEW POTATOES IN MUSTARD SAUCE

1¼ pounds new potatoes
4 tablespoons unsalted butter
1½ tablespoons good-quality Dijon mustard
⅓ cup chopped fresh parsley
Salt
Pepper

In a large pot of boiling water, cook the potatoes until they are tender, about 10 minutes. Drain well and let cool briefly. Peel the potatoes.

In a large skillet over moderate heat, melt the butter. Add the mustard and parsley, and stir. Add the potatoes and toss gently until they are well coated with the mustard mixture. Add salt and pepper to taste, and serve.

Serves 4

PERSIAN SPINACH PIE

1 pound fresh spinach
¾ cup chopped fresh parsley
1 tablespoon chopped fresh dill
2 cups chopped scallions
3 eggs, lightly beaten
½ cup unseasoned bread crumbs
Salt
Pepper
1 tablespoon unsalted butter

Steam the spinach until it is just wilted, about 3 minutes. Drain well, squeezing out as much moisture as possible, then coarsely chop.

In a large mixing bowl, combine the chopped spinach, parsley, dill, scallions, eggs, and bread crumbs. Season to taste with salt and pepper. Mix well.

In a 9-inch skillet over medium heat, melt ½ tablespoon of the butter. When the butter is very hot, add half the vegetable mixture. Spread the mixture evenly in the skillet and cook until the bottom is browned and the eggs are just set, about 5 minutes.

Carefully flip the pie over with a wide spatula. Or, slide the pie onto a flat plate. Hold the plate in one hand and carefully invert the skillet over it. Turn the skillet over again so that the browned portion of the pie is now on top.

Cook for another few minutes until the bottom is browned.

Repeat with the remaining mixture. Cut into wedges and serve.

Serves 4

TABBOULEH

This popular Middle Eastern dish requires bulgur—wheat grains that have been steamed, dried, and then crushed into fine, medium, or coarse "grinds." Once found only in health food stores, bulgur is now quite popular and can easily be purchased in supermarkets. Bulgur is an excellent source of natural fiber, potassium, calcium, iron, and B vitamins.

A true tabbouleh contains bulgur, parsley, olive oil, and lemon juice. Beyond that, anything goes—experiment with whatever fresh vegetables are available.

1 cup medium-grain bulgur
2½ cups water
1 medium onion, finely chopped
6 scallions, finely chopped
1½ cups chopped fresh parsley
½ cup chopped fresh mint (not *dried*)
¼ cup olive oil
¼ cup lemon juice
Salt
Pepper
2 ripe tomatoes, seeded and diced
2 hard-boiled eggs, peeled and sliced

In a medium saucepan, combine the bulgur and water. Cover and cook over low heat until the bulgur has absorbed all the water, about 20 to 25 minutes. Remove from the heat and let stand for 10 minutes.

In a large mixing bowl, combine the bulgur, onion, scallions, parsley, mint, olive oil, and lemon juice, and mix well. Add the tomatoes, and add salt and pepper to taste.

Cover and chill for at least 2 hours. Garnish with the sliced eggs, and serve.

Serves 4 to 6

POTATOES WITH RICE

9 tablespoons unsalted butter
1½ cups long-grain rice
2 cups boiling water
3 medium potatoes, peeled and sliced very thin
1 teaspoon salt
2 garlic cloves, finely chopped
½ cup toasted pine nuts

In a medium saucepan over moderate heat, melt 3 tablespoons of the butter. Add the rice and cook for 1 minute. Add the boiling water, and stir well. Cover, reduce the heat, and simmer for 10 minutes. Drain well.

In a large skillet over low heat, melt the remaining 6 tablespoons of butter. Add the potato slices and the salt, and toss well until the slices are coated with butter. Remove from heat.

Arrange the potato slices evenly in a large, heavy saucepan or Dutch oven. Sprinkle the garlic over the potatoes.

Spoon the cooked rice evenly over the potatoes. Cover tightly and cook over very low heat until the rice is tender, about 45 minutes.

Gently stir the pine nuts into the rice layer, and serve.

Serves 4

SPINACH AND PINTO BEANS

2 tablespoons extra-virgin olive oil
1 teaspoon hot red pepper flakes
1 medium onion, finely chopped
3 garlic cloves, finely chopped
One 15-ounce can pinto beans, drained and rinsed
1 teaspoon balsamic vinegar
Salt
1 pound fresh spinach, coarsely chopped

In a large skillet over moderate heat, heat the olive oil. Add red pepper flakes, onion, and garlic, and sauté, stirring often, until the onion is translucent, about 4 minutes. Add the pinto beans, vinegar, and salt to taste. Cook, stirring often, for 10 minutes. Add the spinach and cook, stirring often, until it is just wilted, about 3 minutes. Serve.

Serves 4

MARINATED BROCCOLI AND CAULIFLOWER

MARINADE:
½ cup extra-virgin olive oil
¼ cup balsamic vinegar
1 teaspoon Dijon mustard
2 garlic cloves, crushed
½ teaspoon dried oregano
½ teaspoon dried basil
½ teaspoon salt
Pepper

1 large head cauliflower, broken into florets
1 large head broccoli, broken into florets
1 red bell pepper, seeded and julienned
1 small onion, thinly sliced

In a mixing bowl, combine the marinade ingredients and whisk well. Set aside.

Bring a large pot of water to a boil over high heat. Add the cauliflower and broccoli pieces and cook until they are just tender, about 3 to 5 minutes. Drain well.

In a large serving bowl, combine the cauliflower, broccoli, red bell pepper, and onion. Pour the marinade over the vegetables and toss well. Cover and refrigerate for 4 hours, tossing occasionally. Remove the garlic cloves before serving.

Serves 4

SALADS

PARSLEY SALAD

5 cups tightly packed parsley leaves
½ cup extra-virgin olive oil
3 garlic cloves, finely chopped
1 tablespoon balsamic vinegar
½ teaspoon salt
Pepper
3 ounces feta cheese
1 cup croutons

Coarsely chop the parsley leaves.

In a medium saucepan over low heat, heat the olive oil. Add the garlic and sauté, stirring often, until the garlic just begins to take on color. Remove from the heat, stir in the parsley, and immediately pour the mixture into a serving bowl.

Stir in the vinegar, salt, and pepper. Stir in the feta cheese and top with the croutons. Serve.

Serves 4 to 6

BLACK BEAN AND PARSLEY SALAD

1 cup finely chopped fresh parsley
One 15-ounce can black beans, drained and rinsed
¼ cup chopped fresh cilantro (coriander)
2 garlic cloves, finely chopped
1 large red bell pepper, seeded and finely chopped
2 celery stalks, finely chopped
4 scallions, thinly sliced
2 jalapeño peppers, seeded and finely chopped
¼ cup lemon juice
3 tablespoons extra-virgin olive oil
½ teaspoon salt
Pepper

In a large serving bowl, combine all the ingredients, and toss well to mix. Marinate for at least 2 hours before serving.

Serves 4 to 6

PARSLEY AND ORZO SALAD

½ pound orzo or any other small pasta shape
1 cup chopped parsley
2 tablespoons chopped fresh mint
1½ cups crumbled feta cheese
1 small red onion, finely chopped
1 large green bell pepper, seeded and finely chopped
⅓ cup extra-virgin olive oil
2 tablespoons lemon juice
Salt
Pepper

Cook the orzo in a large pot of boiling water until it is just tender, about 10 minutes. (Do not overcook.) Drain well and let cool.

In a large serving bowl, combine the orzo and the remaining ingredients, and toss well to mix. Serve immediately.

Serves 4 to 6

POTATO AND EGG SALAD

1 pound red potatoes
4 hard-boiled eggs
1 scallion, thinly sliced
1 celery stalk, coarsely chopped
¼ cup chopped fresh chives
2 tablespoons chopped fresh dill or 2 teaspoons dried dill
1 cup plain lowfat or nonfat yogurt
1 tablespoon Dijon mustard
1 tablespoon fresh lemon juice
1 garlic clove, finely chopped
½ teaspoon sweet paprika
½ teaspoon salt
Freshly ground pepper
2 cups corn kernels

In a large pot of boiling water, cook the potatoes until tender, about 20 to 30 minutes. Drain well. When the potatoes are cool enough to handle, dice them coarsely.

Coarsely dice the hard-boiled eggs.

In large mixing bowl, combine the scallion, celery, chives, dill, yogurt, mustard, lemon juice, garlic, paprika, salt, and pepper to taste. Mix well.

Add the potatoes, eggs, and corn, and toss gently to mix. Cover and chill for 3 to 4 hours before serving.

Serves 4 to 6

SAUCES

BROCCOLI AND ASPARAGUS PESTO

This unusual version of pesto contains no basil. Try it on pasta.

> 2 tablespoons extra-virgin olive oil
> 3 garlic cloves, chopped
> 1 cup walnut pieces
> ½ cup grated Parmesan cheese
> 2 cups chopped cooked broccoli
> 1½ cups chopped cooked asparagus
> ¼ cup milk
> 1 egg yolk
> 1 teaspoon dried oregano
> Salt
> Pepper

Heat the olive oil in a small saucepan over moderate heat. Add the garlic and cook, stirring often, until the garlic is lightly golden, about 3 minutes.

Using a food processor, combine all the ingredients until the mixture forms a smooth, thick paste.

To serve, heat the sauce gently over low heat; do not boil. This sauce stores well in the refrigerator for several days; before storing, carefully pour a thin layer of olive oil over the top to prevent discoloration.

Makes about 2½ cups

PARSLEY VINAIGRETTE

This easy vinaigrette is very versatile. It is an excellent salad dressing, and adds flavor to hot or cold grilled or steamed vegetables.

> ¾ *cup extra-virgin olive oil*
> 2 *tablespoons balsamic vinegar*
> 1 *teaspoon Dijon mustard*
> ¼ *cup chopped red onion or scallions*
> ¼ *cup chopped parsley*
> ½ *teaspoon salt*
> *Pepper*

In a small bowl, whisk together all the ingredients. Use at once.

Makes just over 1 cup

GARLIC AND PARSLEY SAUCE

This sauce is excellent on steamed or lightly sautéed fish.

> 4 *tablespoons unsalted butter*
> 2 *tablespoons finely chopped shallots*
> 2 *garlic cloves, finely chopped*
> 3 *tablespoons finely chopped fresh parsley*

In a small saucepan over moderate heat, melt the butter. Add the shallots and sauté, stirring constantly, until the shallots are softened, about 2 minutes. Add the garlic and parsley and cook, stirring constantly, until the butter turns golden brown. Immediately remove the saucepan from the heat, and serve.

Makes ½ cup

REFERENCES

Aw, T. Y., et al. "Oral Glutathione Enhances Tissue Glutathione In Vivo," *FASEB Journal,* vol. 4, no. 4, abstract 5173, p. A1174, February 1990.

Jones, Dean P., et al. "Content and Bioavailability of Glutathione (GSH) in Food," *FASEB Journal,* vol. 4, no. 4, abstract 5172, p. A1174, February 1990.

Jones, Dean P., Ralph J. Coates, et al. "Glutathione in Foods Listed in the National Cancer Institute's Health Habits and History Food Frequency Questionnaire," *Nutrition and Cancer,* vol. 17, no. 1, pp. 57–75, 1992.

Muldoon, M. F., S. B. Manuck, and K. A. Mathews. "Lowering Cholesterol Concentrations and Mortality: A Quantitative Review of Primary Prevention Trials," *British Medical Journal,* no. 301, pp. 309–14, August 1990.

Oliver, M. F. "Reducing Cholesterol Does Not Reduce Mortality," *Journal of the American College of Cardiology,* vol. 12, no. 3, pp. 814–17, September 1988.

Pennington, Jean A. T. *Bowes & Church's Food Values of Portions Commonly Used,* 16th ed. (Philadelphia: J. B. Lippincott Company, 1994).

TESTING LABORATORIES AND HOLISTIC HEALTH ORGANIZATIONS

MEDICAL TESTING LABORATORIES

Accu-Chem Laboratories
990 North Bower, Suite 800
Richardson, TX 75081
(800) 451-0116

American Medical Testing Laboratories
1 Oakweed Boulevard, Suite 130
Hollywood, FL 38020
(305) 923-2990

Antibody Assay Laboratories
1715 East Wilshire Boulevard, Suite 715
Santa Ana, CA 92705
(800) 522-2611

Diagnos-Tech Clinical and Research Laboratory
6620 South 192nd Place
Kent, WA 98032
(800) 878-3787

Doctor's Data Reference Laboratory
Box 11
West Chicago, IL 60185
(800) 323-2784

Great Smokies Diagnostic Laboratory
63 Zillicoa Street
Asheville, NC 28801
(800) 522-4762

Immuno Laboratories
1620 West Oakland Park Boulevard
Fort Lauderdale, FL 33311
(800) 231-9197

Immunosciences Laboratory
1801 La Cienega Boulevard
Los Angeles, CA 90035
(310) 287-1884

Meridian Valley Clinical Laboratory
24030 132nd Avenue SE
Kent, WA 98042
(800) 234-6825

Meta Metrix Medical Research Laboratory
5000 Peachtree Industrial Boulevard, Suite 110
Norcross, GA 30071
(800) 221-4640

Monroe Medical Research Laboratory
Box 1, Route 17
Southfields, NY 10975
(914) 351-5134

National BioTechnology Laboratory
3212 NE 125th Street, Suite D
Seattle, WA 98125
(800) 846-6285

National Medical Services, Inc.
2300 Stratford Avenue
Willow Grove, PA 19090
(215) 657-4900

Pacific Toxicology Lab
1545 Pontius Avenue
Los Angeles, CA 90025
(310) 479-4911
(800) 32-TOXIC

Physicians' Clinical Laboratories
15613 Bellevue-Redmond Road
Bellevue, WA 98008
(206) 881-2446

Serammune Physicians Lab
1890 Preston White Drive
Reston, VA 22091
(800) 553-5472

WATER TESTING LABORATORIES

National Testing Laboratories, Inc.
6555 Wilson Mills Road
Cleveland, OH 44143
(800) 458-3330

Suburban Water Testing Laboratories, Inc.
4600 Kutztown Road
Temple, PA 19560
(800) 433-6595

HOLISTIC HEALTH ORGANIZATIONS

The organizations listed below can refer you to a medical doctor
or other health care provider trained in holistic health care.

American Academy of Environmental Medicine
4510 West 89th Street
Prairie Village, KS 66207
(913) 642-6062

American Association of Naturopathic Physicians
2355 Eastlake Avenue
Seattle, WA 98102
(206) 323-7610

American Chiropractic Association
1701 Clarendon Boulevard
Arlington, VA 22209
(703) 276-8800

American College for Advancement in Medicine
23121 Verdugo Drive, Suite 204
Laguna Hills, CA 92653
(800) 532-3688
(714) 583-7666

American Holistic Medical Association
4101 Lake Boone Trail, Suite 201
Raleigh, NC 27607
(919) 787-5146

Foundation for the Advancement of Innovative Medicine
  (FAIM)
Two Executive Boulevard, Suite 204
Suffern, NY 10901
(914) 368-9797

SOURCES OF SUPPLEMENTS

Allergy Research Group
400 Preda Street
San Leandro, CA 94577
(800) 782-4274

Biometrics
(800) 724-5566

Douglas Laboratories
600 Boyce Road
Pittsburgh, PA 15205
(888) DOUGLAB

GTC Nutrition Company
Box 843
Broomfield, CO 80038
(800) 522-4682

HealthComm
Gig Harbor, WA 98329
(800) 843-9660

Hickey Chemists
888 Second Avenue
New York, NY 10017
(800) 724-5566

Institute for Rehabilitative Nutrition (IRN)
Box 20199
Columbus Circle Station
New York, NY 10023
(800) 839-6554

Nature's Way Products, Inc.
10 Mountain Springs Parkway
Springville, UT 84663
(801) 489-1520

Phyto-Pharmica
825 Challenger Drive
Green Bay, WI 54311
(800) 553-2370

Primary Nutraceuticals. *See* Institute for Rehabilitative Nutrition.

Solgar Company
500 Willow Tree Road
Leonia, NJ 07605
(201) 944-2311

Thorne Research Laboratories
901 Triangle Drive
Sandpoint, ID 83864
(800) 228-1966

Twin Laboratories, Inc.
Ronkonkoma, NY 11779
(516) 467-3140

Tyson Labs
Wakunga of America Co., Ltd.
23501 Madero
Mission Viejo, CA 92691
(800) 421-2998

Amino acid An organic compound containing oxygen, hydrogen, nitrogen, carbon, and sometimes sulfur. Amino acids are the building blocks of proteins. *See also* Essential amino acids; Nonessential amino acids.

Antioxidant Substances that neutralize destructive free radicals. Glutathione and other antioxidant enzymes are made by your body. Vitamins A, C, and E are antioxidants obtained from your food, as are many other substances such as carotenes found naturally in food.

Carotenes Family of about sixty fat-soluble natural hydrocarbons found in dark green, red, orange, and yellow fruits and vegetables. Some carotenes are converted to vitamin A in the liver; others are antioxidants.

Cysteine Sulfur-containing nonessential amino acid that is one of the building blocks of glutathione.

Dysbiosis Bacterial overgrowth of the small intestine.

Environmental illness (EI) Systemic illness with multiple symptoms caused by overexposure to environmental toxins.

Essential amino acid One of the eleven amino acids available only through foods (see Figure 1.1).

Free radical An unstable, destructive oxygen molecule created as a normal byproduct of metabolism.

Glutamate The ester or salt form of glutamic acid. Essential component of glutathione.

Glutamic acid A nonessential amino acid found in proteins and necessary in the body as a neurotransmitter.

Glutamine A nonessential amino acid closely related to glutamic acid. The most common amino acid in the human body and necessary for a wide variety of important metabolic functions.

Glutathione   A tripeptide made of cysteine, glutamate, and glycine. Found abundantly in the body, it is a powerful antioxidant and detoxifier.

Glutathione peroxidase   A selenium-containing enzyme that catalyzes the reaction between glutathione and free radicals, particularly hydrogen peroxide.

Glutathione reductase   An enzyme that catalyzes the recycling of glutathione within the body.

Glycine   A nonessential amino acid that is one of the building blocks of glutathione.

GSH   Abbreviation for the reduced form of glutathione.

Leaky gut syndrome   Damage to the walls of the small intestine that allows particles of undigested food to enter the bloodstream, possibly causing illness.

Lipid peroxidation   Destruction of the fatty membrane of a cell by free radicals.

Lipoic acid   Coenzyme necessary for recycling of glutathione peroxidase into glutathione reductase; also aids in recycling of other antioxidants.

NAC   Abbreviation for *N*-acetyl-L-cysteine, the manufactured form of cysteine.

Nonessential amino acid   Amino acids manufactured in the body from the essential amino acids.

Peptide   A short, unbranched chain of fifty or fewer amino acids held together by peptide bonds.

Protein   A long chain of more than fifty amino acids arranged in a unique sequence.

Redox operations   Reduction-oxidation processes in the body that neutralize free radicals.

Selenium   A trace mineral necessary as a cofactor for glutathione function.

Silymarin   A derivative of milk thistle used in treating liver disease.

American Journal of Clinical Nutrition, vol. 53 (supplement), January 1991.

Armstrong, D., R. S. Sohal, et al. *Free Radicals in Molecular Biology, Aging, and Disease* (New York: Raven Press, 1984).

Braverman, Eric, and Carl Pfeiffer. *The Healing Nutrients Within: Facts, Findings, and New Research on Amino Acids* (New Canaan, Conn.: Keats Publishing, 1987).

Cooper, Kenneth H., M.D. *Antioxidant Revolution* (Nashville, Tenn.: Thomas Nelson Publishers, 1994).

Dadd, Debra. *Nontoxic, Natural, and Earthwise* (Los Angeles: J. T. Tarcher, 1990).

Erdmann, Robert, Ph.D. *The Amino Revolution* (New York: Fireside Books, 1989).

Golan, Ralph, M.D. *Optimal Wellness* (New York: Ballantine Books, 1995).

Null, Gary. *Reverse the Aging Process Naturally* (New York: Villard Books, 1993).

Passwater Richard A. *Cancer Prevention and Nutritional Therapies* (New Canaan, Conn.: Keats Publishing, 1993).

Passwater, Richard A. *The New Supernutrition* (New York: Pocket Books, 1991).

Passwater, Richard A. *The Antioxidants* (New Canaan, Conn.: Keats Publishing, 1985).

Reuben, Carolyn. *Antioxidants: Your Complete Guide* (Rocklin, Calif.: Prima Publishing, 1995).

Rogers, Sherry A., M.D. *Chemical Sensitivity* (New Canaan, Conn.: Keats Publishing, 1995).

Rogers, Sherry A., M.D. *The E.I. Syndrome* (Syracuse, N.Y.: Prestige Publishing, 1990).

Rogers, Sherry A., M.D. *Tired or Toxic?* (Syracuse, N.Y.: Prestige Publishers, 1990).

Sardi, Bill. *Nutrition and the Eyes* (Montclair, Calif.: Health Spectrum Publishers, 1994).

Shabert, Judy, and Nancy Ehrlich. *The Ultimate Nutrient: Glutamine* (Garden City Park, N.Y.: Avery Publishing Group, 1994).

Steinman, David, and Samuel S. Epstein, M.D. *The Safe Shopper's Bible: A Consumer's Guide to Nontoxic Household Products, Cosmetics, and Food* (New York: Macmillan, 1995).

I N D E X

acetaminophen, 49–51
acetylcholine, 107
acidophilus, 45–46
acne, 164–65
acute bronchitis, 162
adduct, 80
adenosine triphosphate (ATP), 169
adrenal glands, 127
adrenaline, 47, 96
aerobic exercise, 66–67
African Americans, 145, 149
aging, 93–114. *See also* brain; heart
    disease; osteoarthritis
AIDS, 41, 81, 118, 134–35, 168
ajoene, 84
albumin/globulin ratio, 63
alcohol, 29, 41, 48, 140, 141
alfalfa, 129
allergies, 41, 43, 62, 130–31, 159
Allergy Research Group, 47, 106, 113
alpha-lipoic acid, 12–13
alpha tocotrienol, 113
aluminum, 68, 69–70, 107
Alzheimer's Disease, 70, 107–9
American Dental Association, 72
amino acids:
    blood levels of, test for, 18
    cautions about supplements of, 16,
        17–18
    content of, in milk and eggs, 177, 179

    essential, 7
    forms of, 16–18
    functions of, 9
    and levodopa (L-dopa), 109–10
    list of, 7
    nature of, 6–8
    recommended dietary allowances
        (RDAs), 12
    sources of, 8
    supplements of, 16–18
ammonia, 63
amphetamines, 29
angina, 95
anthocyanosides, 146, 147, 148, 150
antibiotics, 41
antibodies, 63, 121–22
anticancer drugs, and lupus, 129
antioxidants 2–5
    and age-related macular
        degeneration, 146
    and aging, 94
    and Alzheimer's Disease, 107
    and arterial damage, 103
    and atherosclerosis, 100–101
    and cancer, 79, 80–82, 84–85, 87,
        88–89
    and environmental illness, 67
    and eye health, 141–43
    and fasting, 35
    and heart attack, 103–4

antioxidants *(continued)*
    and immune system function, 122
    and liver function, 23–25
    and lupus therapy, 129–30
    and osteoarthritis, 111
APOe gene variant, E4, 108
apoproteins, 66, 180
apoptosis theory, 134–35
apples, 75, 159, 176
Applied Nutrition Technology, 132
arginine, 154
arrhythmias of the heart, 104
arsenic, 68, 70, 74
arteries, 95–96, 103
Arthrin, 132
arthritis, 41, 110, 112–13, 130–32
artichoke flour, 46
asparagus, 10, 176
    in recipes, 202
aspirin, 41, 49–51, 129, 159
asthma, 157, 158–62
atheromas, 95, 96–97, 103
atherosclerosis, 95–97, 100–101, 103
autoimmune antibodies, 63
autoimmune diseases, 112, 128–35
avocadoes, 10, 159, 170, 176

bacterial endotoxins, 29, 30
bacterial overgrowth of small
        intestine, 29, 41, 42–47, 49, 52,
        102–3
    confusion of, with environmental
        illness, 62
    and lupus, 129
    and rheumatoid arthritis, 130, 131
    and supplements, 18, 45–47
    symptoms of, 45
    tests for, 51–52
bananas, 170, 176
barberry, 28
barley oil, 113
basophils, 123
B cells, 119–20, 121–22, 129
beans, 11, 75, 177
    in recipes, 198, 200
beer, 159
bell peppers, 130–31
benzene, 57, 58, 65, 66, 80

berries, 159
beta carotenes, 13, 151
BHT (food additive), 65
bifidobacteria, 45–46
bilberries, 147, 150
bile, 23
Bio Detox Powder, 34, 66, 68
Biometrics, 28, 67
biotransformation, 23, 86
    wastes of, 24, 26, 28, 40, 50
black currants, 132
Bland, Jeffrey, 30, 48
blood tests:
    for amino acid levels, 18
    for cholesterol levels, 63
    for environmental illness, 63
    for glutathione levels, 18, 51, 63
    for intestinal permeability, 63
    for kidney and liver enzymes, 63
    for lipid peroxide levels, 51
    for white blood cell count, 63
    for zinc and magnesium levels, 63
blueberries, 150
body fat, ratio to lean mass, 67–68
bone marrow, 120
borage, 132
borage seed oil, 164
bovine cartilage, 111
brain, 104–10
Brazil nuts, 10
breath analysis, 51–52
brewer's yeast, 10, 154
broccoli, 10, 84, 176, 177, 180
    in recipes, 184, 198, 202
bronchitis, 162–64
brussels sprouts, 10, 177
bulgaricus, 45–46
bulgur, in recipes, 196

cabbage, 84, 177
cadmium, 65, 68, 70, 83
caffeine, 35, 37, 49–51
calcium, 68, 74, 96
cancer, 79–91
    curing, 90–91
    and diet, 79, 81, 84–85
    genesis and development of, 80–82
    and liver function, 86–87

reducing risk of, 82–84
sources of information about, 89–90
treatment of, 87–90
Candida albicans (yeast), 43
canola oil, 99
cantaloupe, 178
in recipes, 185, 186
carbon, 6
carcinogens, 80–81, 82–83, 86
carnitine, 100–101, 104
carotenes, 13, 79, 81–82, 124, 141, 164, 167
carotenoids, 145–46
carrots, 13, 129
case histories:
asthma, 157–58
bacterial overgrowth of small intestine, 44
cholesterol, 101
chronic fatigue syndrome, 170
environmental stress, 55–56
immune system and diet, 117–18
liver dysfunction, 22–23
metabolic cleansing, 37–39
old age, 94
osteoarthritis, 112
sluggish liver, 1–2
cashews, 170
catalase, 4, 5, 140
cataracts, 4, 139–41
cauliflower, 176, 177
in recipes, 198
celiac disease, 41
cell-mediated immunity, 121
Centers for Disease Control, 71
cerebrovascular disease, 104–5
chelation, 12, 71, 74
chemical hypersensitivity, 59, 61
chemical messengers, 121, 122, 123
chemotherapy, 87–88
cherries, 159
chicken, 124
chlorine, 64, 83, 103
chlorophyll, 75
chocolate, 154
cholesterol, 66, 96
and atherosclerosis, 97–101

blood levels of, 63, 99
and diet, 98–101
in eggs, 99–100, 179–80
high, 98–100, 101–2, 113–14, 180
and immune system, 126
low, 102–3
and nerve repair, 105
choline, 107
cholinesterase, 57
choroid, 152, 153
chromium, 148, 151
chrondroiten sulfate, 111
chronic bronchitis, 162–64
chronic fatigue syndrome (CFS), 169–70
cigarette smoke, 65, 70, 73, 79, 80, 86, 103, 140, 141, 146, 163, 167
ciliary body, 152
cirrhosis, 17, 48, 171
citrus fruits, 84, 176
cobalamin. *See* vitamins, B_{12}
cofactors, 29
for glutathione synthesis, 6, 10–11, 12–14, 16
cold sores, 17, 153
collagen, 165
collard greens, 81, 145
comprehensive digestive stool analysis (CDSA), 51
conjugation (Phase II detoxification), 23–25, 48, 50, 86
balance with oxidation (Phase I detoxification), 26, 28–30, 62
supplement program for, 29–30
copper, 64–65, 83, 103, 124–25, 160
copper smelting, 70
corn, 13, 129
corneal herpes ulcers, 153–54
corned beef, 159
coronary heart disease, 95–97
costume jewelry, nickel in, 73
coumarins, 84
creatinine, 50
Crohn's disease, 41
cruciferous vegetables, 10, 84, 177
cucumbers, 159
in recipes, 183
cyanohydroxybutene (CHB), 177

CYP1A1, 86
cysteine, 6, 11, 16, 17, 47, 71, 80, 87,
    88, 100, 101, 124, 147, 160–61,
    165, 167, 179
  content of, in selected foods, 178
  foods rich in, 177, 178
cystine, 165
cytochrome P-450 enzymes, 23, 28,
    29, 35, 40, 41, 50, 86

Dadd, Debra, *Nontoxic, Natural, and
    Earthwise*, 64
daidzein, 84–85
dairy products, 131, 164, 177
dandelion, 28
dark green leafy vegetables, 10, 13,
    75, 145, 146
dehydrated soups, 159
deoxyribonucleic acid (DNA), 6, 168
detoxification, and liver function. *See*
    liver
D form amino acids, 16
diabetes, 17, 142, 151–52
diabetic eye disease, 139
diabetic retinopathy, 151–52
diagnostic tests
  for Alzheimer's disease, 108–9
  for bacterial overgrowth in small
    intestine, 51
  for free radical level, 51
  functional liver detoxification
    profile, 49–51, 63
  hair analysis, 73
  for intestinal permeability, 48–49
  for leaky gut syndrome, 48–49, 51
  for liver function, 31–33, 48–52
  provocative chelation (for heavy
    metal detection), 73
  testing laboratories, 205–7
  See also blood tests
diarrhea, 166, 168
diet
  and age-related macular
    degeneration, 145
  and asthma, 159–60, 162
  and bacterial overgrowth of small
    intestine, 44–45
  and cancer, 79, 81, 84–85

and cholesterol, 98–101
cleansing foods, 36, 38–39
and environmental illness, 63, 66–
    67, 68
and glaucoma treatment, 150
and glutathione, 10, 175–203
and heavy metal toxicity, 75
and herpes, 154
and immune system, 117–18, 124
and leaky gut syndrome, 43
and lupus, 129
and multiple sclerosis (MS), 133
and supplements, 8
vs. supplements, 175
Swank, 133
vegan, 131
See also recipes
digestive problems, 166–68
dioxin, 57
discoid lupus, 128
DMSA, 74
dopamine, 109–10
Douglas Labs, 13–14, 25, 28, 67, 126
d-penicillamine, 74
dried fruits, 159
drug abuse, 48
dysbiosis. *See* bacterial overgrowth of
    small intestine

E4 (APOe gene variant), 108
echinacea, 126
eggplant, 130–31
eggs, 11, 66, 75, 105, 107, 124, 147,
    177, 179–81
  amino acid content of, 177
  and cholesterol, 99–100, 179–80
  in recipes, 191, 196, 201
egg yolks, 13
endotoxins, 41
environmental illness, 55–75
  blood tests for, 63
  confusion with other illnesses, 61–
    62
  curing, 64–68
  and diet, 63, 66–67, 68
  and eggs, 180
  supplements for, 67–68
  susceptibility to, 62–63

symptoms of, 59–61
See also heavy metals
Environmental Protection Agency
(EPA), 58
environmental toxins, 56–58, 59, 63
and Alzheimer's Disease, 107
and arterial damage, 103
and asthma, 160, 162
in the home, 64–65
and liver function, 24, 40
reducing exposure to, 82–83
self-quiz, 26–27
in the workplace, 65–66
eosinophils, 123
EPA (fatty acid), 164
epilepsy, 17
Epstein, Samuel S. (with Steinman),
The Safe Shopper's Bible, 64
estrogen, 47, 85
evening primrose, 132
exercise, and lymphatic system, 120
eyes, 4, 139–54
supplements for, 141–44
See also specific eye diseases

fabric softeners, 57, 58, 64, 72
fasting, 35, 126–27, 131
fatty acids, and immune system, 126
fibrinogen, 96–97
fillings in teeth, mercury in, 72
fish, 124, 147, 177
fish oils, 99, 111, 129, 132, 150, 151,
164
flavonoids (phytochemicals), 84, 141
flaxseed oil, 99, 150, 164
fluid intake, 25, 35, 36, 37, 66, 75,
119, 127, 165–66
foam cells, 96
foam rubber, 57, 64
folic acid, 107
foods
as asthma triggers, 159
cysteine-rich, 177, 178
glutathione-rich, 10, 176–77
xenobiotic, 65
See also diet; recipes
formaldehyde, 57, 63, 65
formic acid, levels in blood, 63

free-form amino acids, 16, 17, 89,
127, 132, 170
free radicals, 2–5, 40, 41, 86, 131
and aging, 93–94
and Alzheimer's Disease, 107
and arterial damage, 103
and asthma, 160
and cancer, 80–81, 88
destruction of fatty acids by, 5
and heart attack, 104
and immune system function, 125
and inflammation, 153, 167
and iron, 83
and LDL oxidation, 98
and macrophages, 122
and osteoarthritis, 111
and oxidation, 23–24, 26, 28
and psoriasis, 164
and rheumatoid arthritis, 130
and synovial fluid sacs, 130
test for levels of, 51
and ultra-violet light, 139–40, 146
fructooligosaccharides (FOS), 44, 46–
47
fruits, high-pectin, 75
functional liver detoxification profile,
49–51, 63

gamma linoleic acid (GLA), 132,
164
garlic, 46, 74, 84, 99, 126
in recipes, 203
gastrointestinal tract, and glutamine,
168
gelatin, 154
genestein, 84–85, 113–14
genetic damage, 80
Germano, Carl, 11
gerontology, 93
ginkgo biloba, 104–5, 106, 107, 147
glaucoma, 139, 148–50
globulin, blood levels of, 63
glossary, 213–14
glucosamine sulfate, 111
glutamate, 168
glutamic acid, 6, 11, 168, 179
glutamine, 11, 17, 42, 44, 47, 89, 106,
167–68

glutamine synthetase, 108
Glutaplex, 14, 25, 67, 165
glutathione:
  and Alzheimer's Disease, 107
  as antioxidant, 4–6, 24–25
  and arsenic, 68
  and arthritis therapy, 111
  blood tests for levels of, 18, 51
  boosting levels of, 10–11
  cofactors for synthesis of, 6, 10–11, 12–14, 16
  components of, 5–6
  and conjugation (Phase II detoxification), 24–25
  content of, in selected foods, 176
  dietary sources of, 10, 176–77
  diet vs. supplements to maintain levels of, 8
  discovery of, 4
  enzymes necessary for, 5
  forms of, 5
  levels of, in humans vs. other mammals, 94
  overview of benefits of, 1–3, 4–5
  oxidized form (GSSG), 5
  plant foods that stimulate, 177–79
  and radiation therapy, 8
  reduced form (GSH), 5
  and sulfation, 47–48
  time frame of improvements from supplements of, 19
glutathione peroxidase, 4, 5, 13, 62
glutathione reductase, 5, 6
glycine, 6, 11, 179
goldenseal, 28
googul and Googulplex, 114
grains, 10, 11, 108, 124, 177
granulocytes, 121, 122–23
grapefruit, 10
grapefruit juice, 29
grapes, 159
green peppers, 159
GSH 250 Master Glutathione Formula, 13–14, 16, 28, 34, 35, 36–37, 43, 67, 68, 86, 118, 126, 163, 164–65
Gulf War syndrome, 56–57

hair, 165
hair analysis, 73
healing of wounds, 157, 168
HealthComm, 34, 48
heart attack, 95, 103–4
heart disease, 95–104
heavy metals, 29, 57, 86
  and Alzheimer's Disease, 107
  and arterial damage, 103
  and asthma, 160, 162
  and conjugation (Phase II detoxification), 24
  environmental sources of, 69
  poisoning by, 63, 68–73
  supplements for toxicity of, 74–75
Helicobacter pylori, 166–67
helper T cells (T-4 cells), 121, 122, 134
heparin, 123
hepatitis, 48, 170–71
Hepatox, 27–28, 67
herbicides, 40, 64, 65, 70
herpes, 17
herpes simplex eye infections, 139, 153–54
Hickey, Jerry, 114
Hickey Chemists, 114
high blood pressure, 103
high cholesterol, 98–100, 101–2, 113–14, 180
high-density lipoproteins (HDLs), 97–101, 102
high-fat diet, 29, 126
histamine, 123
HIV, 41, 81, 118, 134–35
holistic health organizations, 207–8
homocysteine, 100
honey, 46
hormones, stress-related, 47, 127–28
human immunodeficiency virus. See HIV
humidifiers, 119
hydrocarbons, 80, 103
hydrogen, 6, 52
hydrogen peroxide, 4, 5, 122
hydroxyl radical, 4
hypocholesterolemia, 103
hypothalamus, 127

iberin, 177
ibuprofen, 41
IgA antibodies, 122
IgE antibodies, 122
IgG antibodies, 121–22
immune system, 117–35
  boosting, 123–27
  and diet, 117–18, 124, 126
  diseases of, 128–35
  and glutamine, 168
  how it works, 118–23
  and stress, 127–28
immunoglobulins (Ig), 121–22
indoles, 84
inflammatory bowel disease, 41
initiation (cancer genesis), 80
insecticides, 57, 64, 65
Institute of Rehabilitative Nutrition,
  29, 30, 34
insulin, 17
interferon, 123
interleukin, 63, 123
intestinal permeability test, 48–49,
  63
iritis, 152–53
iron, 29, 65, 83, 160
isoflavones (phytoestrogens), 84–85,
  113–14
isoleucine, 169
isothiocyanates, 84

kale, 13, 75
keratin, 165
kidney disease, 17
kidney failure, 95
kidneys, 23, 63
kidney stones, 171
killer (cytotoxic) T cells, 121
Kyolic, 74, 126

lactobacillus, 34, 45–46
Lactobacillus bulgarica, 46
lactulose, 49
latex house paint, 72–73
L-canavanine, and lupus, 129
lead, 64–65, 68, 70–72, 74, 83, 124–
  25, 160
leaky gut syndrome, 28, 29, 40–43, 52

  and asthma, 161
  confusion of, with environmental
    illness, 62
  and nonsteroidal anti-inflammatory
    drugs (NSAIDs)s, 111–12
  and rheumatoid arthritis, 130, 131
  symptoms of, 42
  tests for, 48–49, 51
lecithin, 107
lentils, in recipes, 194
leucine, 169
leukemia, 58, 80
leukocytes (white blood cells), 63, 81,
  120–23, 125
levodopa (L-dopa), 109–10
L form amino acids, 16
lipid peroxidation, 5
lipid peroxide, 51
lipoic acid, 6, 10, 12–13, 16, 101,
  130, 135
lipoproteins, 66
liquid diet, 34–36, 131
liquid fasting, 35
liver, 21–52
  alcoholic damage to, 48
  and antioxidants, 23–25
  blood tests for enzymes of, 63
  and cancer, 86–87
  and cholesterol, 102
  drug-abuse damage to, 48
  dysfunction in, 1–2, 21–23, 30, 40–
    48, 62, 102, 110, 164
  functioning of, 12, 23–24
  functions of, 21
  glutathione supplements for
    diseased, cautions with, 17
  imbalances in functions of, reasons
    for, 40–48
  Parkinson's Disease and function
    of, 110
  and stress, 47–48
  supplements for healthy, 25–26, 28–
    30
  tests for level of functioning of, 31–
    33, 48–52
  up-regulation of, 26–30, 40, 51
lovastatin (Mevacor), 98
low cholesterol, 102–3

low-density lipoproteins (LDLs), 97–101, 102, 113, 114, 180
lupus erythematosus, 128–30
lutein, 13, 145
lycopene, 13, 84
lymphatic system, 119–20
lymph nodes, 120, 124
lymphocytes, 63, 121–22, 126
lymphokines, 123
lysine, 17, 125, 154

macrophages, 119, 122, 123, 124, 125, 126, 163, 167
macular degeneration, age-related (AMD), 4, 139, 144–48
magnesium, 29, 62–63, 68, 74, 100, 141, 150, 151–52, 169
malabsorption problems, and amino acid supplements, 18
malignant melanoma, 83–84
manganese, 108
mannitol, 49
Marplan, 18
mast cells, 123
meats, 10, 11, 124, 147, 154, 177
melons, 159, 176
memory loss, 106, 109
mercury, 68, 72–73, 74
metabolic cleansing, 25–26, 30–39, 66, 68, 102, 127, 164
  and arthritis therapy, 113
  and asthma, 162
  and cancer prevention, 86–87
  and chronic fatigue syndrome, 169
  and Parkinson's Disease, 110
  supplements during, 34–35, 36, 39
methane gas, in breath analysis, 52
methionine, 6, 100, 110, 125
milk, 10, 154, 170
  amino acid content of, 177
milk thistle plant (*Silybum marianum*), 27
Mitochondrial Resuscitate, 48
molybdenum, 29
monoamine oxidase (MAO) inhibitor drug, 18
monocytes, 96, 121, 122
monokines, 123

monosodium glutamate (MSG), 159
monounsaturated fats, 99, 180
moth balls, 64
mucus membranes, 118–19, 120
multiple chemical sensitivity, 61
multiple sclerosis (MS), 132–34
multivitamins, 63, 75

NAC, 11, 16, 17, 67, 86, 88, 101, 130, 134–35, 141, 147, 152, 160–61, 163–64, 171
N-acetyl-L-cysteine capsules. *See* NAC
Nardil, 18
naringinen, 29
National Cancer Institute, 84
natrol, 105
natural killer (NK) cells, 121, 122
Nature's Plus, 89, 142
Nature's Way, 63, 75, 161
neovascularization, 151
nerve cell growth, 12
neurotoxins, 110
neurotransmitter chemicals, 106, 107
neutrophils, 122–23, 126, 167
niacin, 29
nickel, 68, 73
nightshade family, 130–31
nitrogen, 6
nitrogen-based wastes, 8
nonsteroidal anti-inflammatory drugs (NSAIDs), 41, 111–12, 153, 159, 166
nuts, 124, 154

Occupational Safety and Health Administration (OSHA), 71
Ocucare, 142
Ocugard, 142
okra, 176
olive oil, 99
olives, 159
omega-3 fatty acids, 99, 111, 129, 132, 150, 151
onions, 84, 99
oral contraceptives, 124
organ meats, 10
organophosphates, 57
Ornish program, 98–99

osteoarthritis, 110–13
osteoporosis, 43, 69
oxidation, 4
oxidation (Phase I detoxification),
    23, 24, 48, 50–51, 86
    balance with conjugation (Phase II
        detoxification), 26, 28–30, 62
    supplement program for, 29–30
oxidative stress, 26, 30, 35, 43, 51, 63,
    160, 162, 168
oxidative stress panel, 51
oxygen, 6
oysters, 124

palm oil, 113
Parkinson's Disease, 109–10
parsley (*Petroselinum crispum*), 10, 178–
    79, 180
    in recipes, 181, 182, 196, 199, 200,
        203
Passwater, Richard A., 12
pathological detoxifiers, 28–29
peaches, 159, 176
peanuts, 170
pears, 75, 176
pepsin, 166
pesticides, 40, 64, 70
Phase I supplement program, 30
Phase II supplement program, 29–
    30
phenol-based compounds, 47
phenylalanine, 18, 125, 169
phenylketonuria, 18
phosphatidyl serine (PS), 106
phytochemicals (flavonoids), 84, 141
phytoestrogens (isoflavones), 84–85,
    113–14
Phytogreens, 75, 85
Phytopharmica, 114
pineapple, 108
pink bismuth (Pepto-Bismol), 166
pituitary gland, 127
plant foods, glutathione stimulation
    by, 177–79
plaque, 96–97
platelets, 9, 96–97, 100, 114
plums, 159
pollen, 159, 160

polycyclic aromatic hydrocarbons
    (PAHs), 86
polyester, 57
polyunsaturated fats, 180
potatoes, 10, 130–31, 159, 176
    in recipes, 184, 186, 187, 189, 190,
        191, 192, 193, 194, 197, 201
poultry, 10, 177
pregnancy, 124
premenstrual syndrome (PMS), 21
preservatives, 65
Primary Nutraceuticals, 43, 46, 106
probucol, 98
Prodophilus Complex, 46
prostaglandins, 47, 111, 132, 165
proteins, 6–8
provocative chelation test, 73
psoralens, 179
psoriasis, 41, 157, 164
psychoneuroimmunology, 127
psyllium fiber, 34
Pure L-Glutamine-Intestinal
    Permeability Powder, 43
pyridostigmine, 57
pyridoxine. *See* vitamins, B_6

Q-10 co-enzyme, 98
quercetin, 30, 84, 147

radiation therapy, 87, 88, 129
raisins, 108
recipes, 180–203. *See also* salads;
    sauces; side dishes; soups
recommended dietary allowances
    (RDAs):
    amino acids, 12
    for overall health, 15
red blood cells, 120
redox operations, 4, 5
reduction, 4
Reye's syndrome, 17
rheumatoid arthritis (RA), 41, 110,
    111, 130–32
riboflavin. *See* vitamins, B_2
rice, in recipes, 197
rice bran, 113
Rogers, Sherry A., 61
root beer, 159

salads, 199–201
  black bean and parsley salad, 200
  parsley and orzo salad, 200–201
  parsley salad, 199
  potato and egg salad, 201
salicylates, 159
salmon, 10
salmonella, 181
Sardi, Bill, 143
saturated fats, 180
sauces, 202–3
  broccoli and asparagus pesto, 202
  garlic and parsley sauce, 203
  parsley vinaigrette, 203
schizandra (wuweizi), 28
seafood, 10
selenium, 6, 10, 13, 62–63, 68, 79–80,
    82, 100, 111, 124, 130, 141, 150,
    152, 160, 164
self-quizzes:
  exposure to environmental toxins,
    27
  liver function, 31–33
shark cartilage, 111
shrimp, 159
sick building syndrome, 65, 82–83
side dishes, 189–99
  broccoli and cauliflower,
    marinated, 198–99
  eggs, curried, and potatoes, 191
  lentils, curried, with spinach, 194
  potato, mashed, pie with cheese,
    189
  potato and zucchini frittata, 193
  potatoes, Indian smothered, 192
  potatoes, new, in mustard sauce,
    194–95
  potatoes with rice, 197
  potato gnocchi, 190
  spinach and pinto beans, 198
  spinach pie, Persian, 195
  tabbouleh, 196
silver amalgam fillings, 72
silymarin, 26–28, 48, 171
Sinemet, 110
singlet oxygen, 4
skin, 83–84, 118–19, 164–66
sluggish liver, 1–2, 30

and cholesterol levels, 102
confusion of, with environmental
    illness, 62
and Parkinson's Disease, 110
and psoriasis, 164
sodium nitrate (additive), 65
Solgar Company, 11, 13, 14, 63, 73,
    75, 85, 89, 113, 114, 161
sotanine, 131
soups, 181–88
  broccoli and potato soup, 184
  cantaloupe and tomato soup,
    chilled, 185
  cucumber and spinach soup,
    chilled, 183
  parsley soup, 181
  parsley with tarragon soup, chilled,
    182
  potato and cantaloupe soup,
    Mexican, chilled, 186
  spinach and potato soup, 187
  spinach and yogurt soup, with
    fresh herbs, 188
soybeans, 84–85, 113, 129, 154
spinach, 10, 108, 129, 145, 176, 178,
    180
  in recipes, 183, 187, 188, 194, 195,
    198
spleen, 120
split peas, 124
spreading phenomenon (multiple
    chemical sensitivity), 57
squash, 176
Steinman, David (with Epstein), *The
    Safe Shopper's Bible*, 64
stem cells, 120
steroid hormones, 127
steroids, 41, 142
strawberries, 10, 176
stress:
  and cancer, 87
  and immune system, 127–28
  and liver function, 47–48
stroke, 95, 104–5
sugar, and white blood cells, 126
sulfation, 47–48, 50, 87
sulfhydryl, 6
sulfites, 159

sulforaphanes, 84, 177
sulfur, 6, 179
sulfur dioxide, 57
sunglasses, 141, 143, 146, 154, 166
sunlight, and skin cancer, 83–84
superoxide dismutase (SOD), 4, 5, 122, 124, 140
superoxide radical, 4, 5
supplements:
   for age-related macular degeneration, 146–48
   for AIDS, 134–35
   for Alzheimer's Disease, 108
   amino acid, 16–18
   for arthritis, 112–13
   for asthma, 160–62
   and atherosclerosis, 100–101
   building blocks of glutathione, 11
   and cancer, 88–89
   for cancer, 82, 84–85, 86–87
   cautions about amino acid, 17–18
   and cerebrovascular disease, 104–5
   for chronic fatigue syndrome, 170
   for conjugation (Phase II detoxification), 29–30
   controversy about glutathione supplements, 10
   daily RDA's for basic health, 15
   for diabetes, 151–52
   and diet, 8
   vs. diet, 175
   for environmental illness, 67–68
   for eye health, 141–44
   for glaucoma, 150
   for heavy metal poisoning, 74–75
   for hepatitis, 171
   for herpes, 154
   for high cholesterol, 101–2
   for immune system, 117–18, 124–27
   for leaky gut syndrome, 43
   for liver health, 25–26, 28–30
   for lupus, 129–30
   and memory loss, 106
   during metabolic cleansing program, 34–35, 36, 39
   for multiple sclerosis (MS), 133–34
   need for, 8
   for osteoarthritis, 111–13
   for psoriasis, 164
   to reestablish intestinal bacteria, 18
   for rheumatoid arthritis (RA), 131–32
   sources (retail) for, 209–11
   for stress, 48
   to reestablish intestinal bacteria, 45–47
   for ulcers, 167–68
suppressor T cells, 121, 129
surgery, 87
Swank diet, 133
sweating, 67
sweet potatoes, 13
Swiss chard, 75
systemic lupus erythematosus (SLE), 128–30

T-4 cells (helper T cells), 121, 122, 134
taurine, 7, 104, 147, 148, 152, 169
T cells, 119, 120, 121, 122, 124–25, 127, 134
tea, 108, 159
testosterone, 47
thiamine. See vitamins, B_1
thymus, 119, 120, 121, 124, 125, 128
tocotrienols, 113
tomatoes, 13, 84, 130–31, 159, 176
   in recipes, 185, 196
total protein, blood levels of, 63
toxins in the environment. See environmental toxins
triglycerides, and immune system, 126
tripeptides, 5–6
tryptophan, 18, 110, 125, 169
tumor necrosis factor (TNF), 135
tumors, 81–82
Twin Labs, 14, 89
Twinlife, 142
tyrosine, 18, 110

ulcerative colitis, 168
ulcers, 157, 166–68
Ultimate Bionetics, 63
Ultra Clear, 34, 66, 68

ultra-violet (UV) light, 83–84, 139–40, 145–46, 154, 166
up-regulation, of liver function, 26–30, 40, 51
uric acid, 8
urinary amino acid screening, 18
uveitis, 139, 152, 153

vasoconstrictors, 123
vegan diet, 131
vitamins:
  A, 13, 79, 81, 111, 124, 141
  B$_1$ (thiamine), 48
  B$_{12}$ (cobalamin), 107, 124
  B$_2$ (riboflavin), 6, 10, 13, 29, 48
  B$_6$ (pyridoxine), 100, 124, 147, 151
  B-complex, 48, 106, 107, 124, 141
  C, 12, 13, 74, 79, 81, 100, 111, 124, 135, 140, 141, 142, 150, 154, 160, 163, 164, 167, 171
  D, 97, 180
  E, 12, 13, 79, 81, 82, 100, 101, 103, 111, 113, 124, 129, 141, 142, 150, 164, 167

walnuts, 176
water:
  cadmium in, 65, 83
  chlorinated vs. filtered, 83
  chlorine in, 64, 83
  copper in, 64–65, 83
  iron in, 65, 83
  lead in, 64–65, 83
  purifiers and testers of, 64–65
water intake. *See* fluid intake
watermelons, 10, 13, 84, 178
Water Pure filter system, 65, 83
water testing laboratories, 207
wheat germ, 170
Whitaker, Julian, *Reversing Diabetes*, 151–52
white blood cells (leukocytes), 63, 81, 120–23, 125
wine, 159
wuweizi (schizandra), 28

xenobiotics. *See* environmental toxins

yarrow, 28
yeast, 42, 43, 103, 129, 130
yeast (as food), 164
yogurt, 45
  in recipes, 188

zeaxanthin, 13, 145
zinc, 6, 13, 62–63, 74, 106, 124, 141, 150, 152, 154, 164
zinc citrate, 124
zinc picolinate, 124
zinc refining, 70
zucchini, in recipes, 193